Sri Garib Dass Oriental Series NO. - 158

LOGIC OF KNOWLEDGE BASE

(A Nyāya Reader for Designing Computational Lexicon)

DR. KESHAB CHANDRA DASH
Reader & Head
Dept. of Nyāya Philosophy
Sri Jagannath Sanskrit
Vishvavidyalay
PURI- 752002

Sri Satguru Publications
A Division of
INDIAN BOOKS CENTRE
DELHI - (INDIA)

Published by :
Sri Satguru Publications
Indological and Oriental Publishers
A Division of
Indian Books Centre
40/5,Shakti Nagar,
Delhi-110007.
(INDIA)

First Edition : Delhi, 1992

ISBN 81-7030-346-X

Printed at :
D.K. Fine Art Press, Delhi.

PRINTED IN INDIA

PREFACE

A thought assumes an inherent correspondence between a linguistic form and the object referred to by it. A linguistic form thus becomes a base which has to be a content of consciousness. The form and content shape themselves into revelation and realisation which, in other words, become knowledge-base and unit of reference respectively when viewed through modern perceptions.

This work, in its schematic perspective, presents information resources about the structure of a lexicon. It provides a clue to the issue of designing computational lexicon for natural language analysis system. Of course, there is a choice of selecting criteria which incorporates theoretical and subjective predilections for lexicon designers. But this attempt involves a sensibility of rediscovering the Nyāya model which specifics the nature of items associated with a word. As far as the point of lexical recognition in Indian languages is concerned the Nyāya model would be of immense help for determining the semantic properties of a word. Further, some possible storage schemes may be designed and developed on the basis of the theoretical efficiency of this model for operation in computer.

The text consists of the portions from four source texts namely *Nyāya-bhāsya* of Vātsyāyana (4th cent. A.D.), *Nyāya-mañjarī* of Jayanta Baṭṭa (10th cent. A.D.), *Tarka-bhāṣā* of Keśava Mishra (13th cent. A.D.) *Śabda - śakti - prakāśikā* of Jagadiśa Tarkālaṁkāra (17th cent. A.D.).

Although the arrangement of topics involves the problem of temporal differences a separate historical outline is not necessary as the framework of thought maintains the theoretical balance. The idea of inclusion of original texts is to ensure consistency of a well-defined and developed system as a whole and to show its relevance to the issues that determine the aspects of a word expert system.

The topics of the text may be broadly divided into two major sections such as (i) category information and (ii) subcategory information. The first deals with the aspects of categories enumerated in Nyāya system, the concept of meaning, the relations that subsists between a word and its meaning, the referent of a word, manifoldness of meaning, valid word and aspects of nominals. The second which deals with sub-category information points to a fact that the concept structure of a word ultimately resorts to a referential setup in which its inherent features are determined.

As this volume stems out from the inspirations of my well-wishers I owe my gratitude to all those who enrich my interest.

I express my feelings of indebtedness to Professor V.N. Jha, Director, Centre of Advanced Study in Sanskrit, University of Poona, Pune, Professor P.K. Mukhopadhyaya, Professor of Philosophy, Centre of Advanced Study in Philosophy, Deptt. of Philosophy, Jadavpur University, Calcutta, Professor Radha Vallabh Tripathy, Professor & Head, Deptt. of Sanskrit, Dr. H.S. Gour University, Sagar (M.P.) who always encourage me in different ways for higher and deeper learning.

I acknowledge my deepest gratitude to Professor G.C.

Nayak, Vice Chancellor and Shri S.K. Dash Chaudhury, Registrar and Comproller of Finance, Shri Jagannath Sanskrit Vishvavidyalaya, Puri, whose inspiration has given me a courage to walk ahead with conviction in my capabilities.

At this moment, beyond all logics and formalities I enjoy a feeling of admiration for my wife Smt. Subhadra Dash for her untutored co-operation in all my literary and scholastic endeavours.

Keshab Chandra Dash

Puri

CONTENTS

INTRODUCTION

Language understanding has become the main topic in theory-oriented research of computational linguistics. This is also gaining its identity in the field of Artificial Intelligence researches. The basic idea of modelling language understanding is to represent the knowledge structure of what holds about the situations, objects, actions and their relations etc. that are referred to by words or described by the sentence or texts. To this effects, AI approach consists in the integration of language information. The term information, in other words, points to the universe of knowables which consists of various concepts, ideas, facts, names and so on. This is an apparent impasse between communicables and communications, i.e., what one wants to communicate and the way in which it is communicated. This process, in its wider sense, referes to nothing but 'meaning' which is resolved in the form of content and representation. AI approach aims at the development of methods of qualitative combination of content and representation for simulation on computer.

The relational representation is a kind of internal representation. This is considered from two different points of view. The first is descriptive, which deals with static representation of the information embedded in the text.

The second one is considered from processing point of view. This deals with the way in which the information is assimilated by the system. In other words, it shows how the information is incorporated with existing knowledge structures and organised in the memory. Thus while designing a representational model of a language, one has to take note of the types of information encoded in a text. At first hand, the levels of representation are delimited, i.e., the relevant units of analysis are distinguished such as : 1. lexical items, 2. the clause, 3. the sentence and the text as a whole.

Knowledge Base :

The area of AI research is well known as expert system or knowledge based system. The expert system points to a computer programme which contains both declarative knowledge and procedural knowledge. The declarative knowledge consists of the facts about objects, events and situations etc. whereas the procedural knowledge takes recourse to the information about action. These two types of expert systems are rule-based and model-based. By rule based system the procedural knowledge is integrated with the declarative knowledge in the form of production rules. Similarly the model based system is based on a model of the structure and behaviour of the device.

When viewed on above lines the knowledge base is technically known as the component of expert system that contains the system's declartive and procedural knowledge. Further, the knowledge base is divided into three major categories, namely context, parameters and production rules. A context refers to a problem area in the domain of expert system. It provides a convenient way of dividing the larger problem into smaller to meet a specific goal. The parameter is associated with system's conclusion, i.e. the user of the system supplies the values of the parameters in response to questions or problems. The relationship among values of parameters appear in the form of production rules. In other words, the inference mechanism is encoded in the form of production rules which specify the logical relationship of the values of parameters. The production rule gives ability to the inference engine to infer the values of the facts that appear as conclusions.

Nyāya Model And Representation :

On the basis of above discussion we may consider here the Nyāya model of representation from descriptive point of view. Here, we take the lexical representation level. The lexicon has two main functions such as : first, it shows its grammatical dependency and secondly, it records some of its semantic properties. From Nyāya point of view a lexicon is judged as a base containing certain knowledge which is expressed in terms of its predicability. In other words, it is a meaningful unit (*śaktam padam*). In the present context, the base could be rightly termed as a knowledge base. The Nyāya system provides certain principles for its representation.

When we talk of computational lexicon it is a fact that the type of information required to be stored in the lexicons of natural language processing systems should be described. The informations are devided into two major categories namely (1) intra-word information and (2) inter-word information. The intra-word information pertains to individual word only. Inter-word information points to the way in which a word is related to other word.

Intra-Word Information :

From Nyāya point of view, the structure of a lexicon which provides intra-word information may be presented in the following diagram.

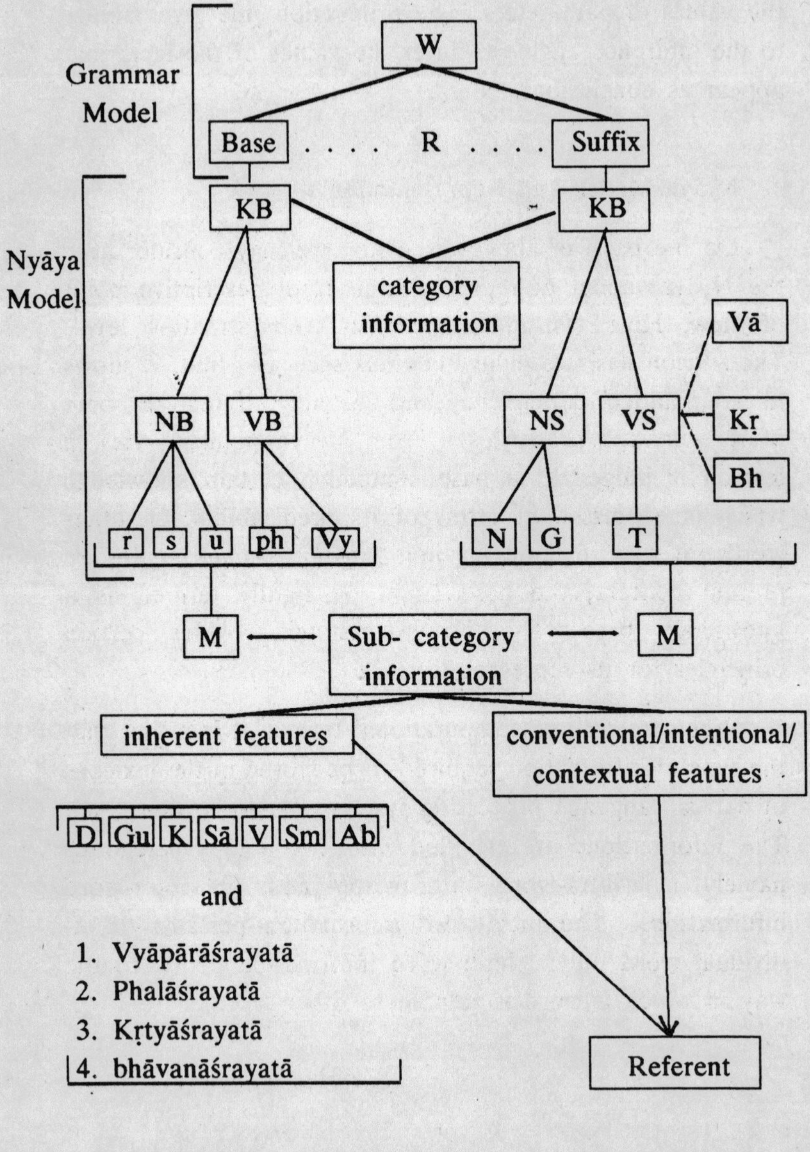

W = word	NS = Nominal Suffix	Ph = Phala (result)
R = Relation	VS = Verbal Suffix	Vy = Vyāpāra (action)
KB = Knowledge Base	I = Individual	N = Number
NB = Nominal Base	S = Shape	G = Gender
VB = Verbal Base	U = Universal	T = Tense
		M = Meaning

Va = Vyāpārāśrayatā (Resorting ot action)

Kṛ = Kṛti (Volitional effort)

Bh = Bhāvanā (productive urge)

D = Dravya (substance)	Ab = Abhāva (Absence)
Gu = Guṇa (quality)	££££ = Relation
K = Karma (action)	—— = alternative
Sā = Sāmānya (universal)	Predominant element
V = Viśeṣa (particular)	⟵⟶ = Relational State.
Sm = Samavāya (inherence)	

This analysis helps in selecting the information by way of specifying the nature of items required for processing.

We can examine an abstract word *īśvara* 'God' taking recourse to the above model in the following manner :

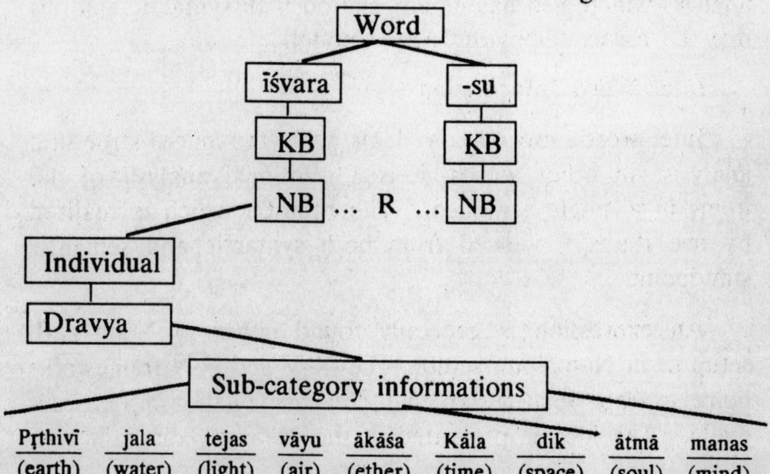

This shows that the owrd *īśvara* ultimately refers to 'soul'.

The general model of processing computational lexicon is considered from the standpoint of parser. For instance,'one to one' relational pattern is presented in the following mannar.

Word → [Morphological analysis] → [Syntatic analysis] →[Semantic analysis] Similarly, 'one to many rlation' between word and its category is presented below for the word INTIMATE.

[ADJ : base : INTIMATE form : positive]	→ [ADJ : INTIMATE takes prep : WITH]	→ [semantics for INTIMATE-ADJ]
[V : INTIMATE tense : Present: number : Pl]		
[V: base : INTIMATE form : infinitive]	→ [V : INTIMATE complements THAT COMP]	→ [Semantics for INTIMATE -V]
[V : Base : INTIMATE tense : Present, Per : 1 Number : sg.]		

A parser ultimately associates each word in full understanding system.

As far as the Indian languages are concerned the morphological and syntactic analysis may be considered from the standpoint of Grammar (Pāṇinian) model. The semantic analysis which also has its correlation with syntactic analysis may be associated with Nyāya model.

Inter-Word Information :

Inter-word information deals with the micro-structural analysis. In other words, it is a relational analysis of the items in a single sentence. The relation which is realised by two items is viewed from both syntactic and semantic standpoint.

An expression is generally found either in Noun-Verb setup or in Non-Noun setup. The N–V and N-N frameworks point to case relationship and non-case relationship respectively. The idea is presented in the following diagram : -

1. Noun-Verb frame Work :

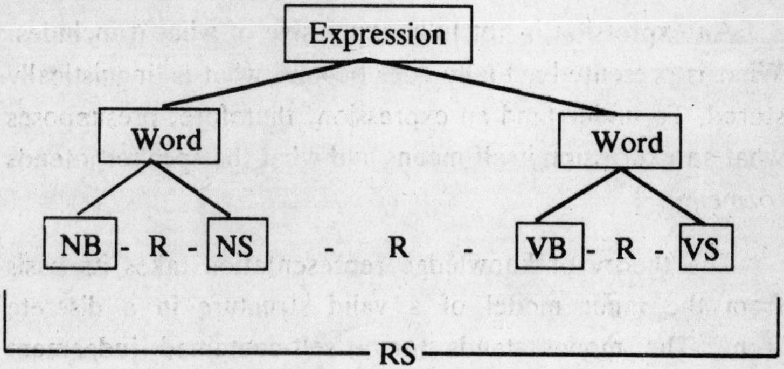

2. Noun - Noun frame work :

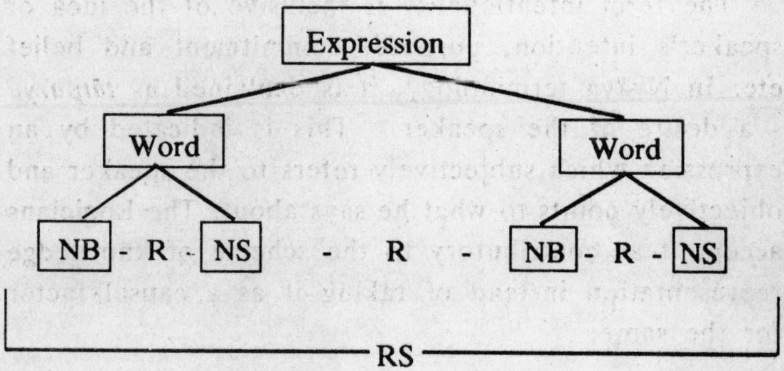

RS = Relational Structure.

Such a mechanism requires contents of the structure as well as extralinguistic and situational features. This has been discussed in detail elsewhere in my book 'Relations in Knowledge Representation.'

Moreover, it may be said that the above model can be satisfactorily implemented by considering the nature of relation and the relata as per requirement of the system.

Intentionality :

An expression is not fully expressive of what it includes. What is presented actually goes beyond what is linguistically stored. To understand an expression, therefore, presupposes what an expression itself means and what the speaker intends to mean.

The theory of knowledge representation takes its basis from the inner model of a valid structure in a discrete form. The model stands for a self-contained judgement which is mostly constrained by the extra linguistic factors, like context, convention and intentionality etc.

The term intentionality is inclusive of the idea of speaker's intention, purpose, commitment and belief etc. In Nyāya terminology, it is explained as *tātparya* - a desire of the speaker. This is indicated by an expression which subjectively refers to the speaker and objectively points to what he says about. The Logicians accept it as contributory to the scheme of knowledge representation instead of taking it as a causal factor for the same.

***Tātaparya* as Desire :**

The form of tātparya is "X should mean Y" (X stands for expression and Y for meaning). Tātparya is related to the speaker by the relation 'inherence'. X and Y are related to the speaker by a relation called *viṣayatā* contentness. *'The knowledge produced in the addresse is qualified by the relation uccāritatva* 'utterability'. The desire points to both speaker and addressee. This is "intention" in case of the speaker and "understandability" in case of the addressee.

The whole idea may be put in the following diagram :

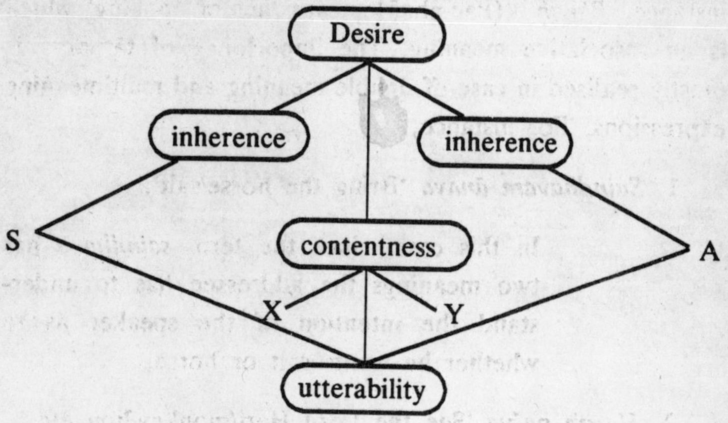

S = Speaker

A = Addressee

X = Expression/Word

Y = Meaning

Straight line = Direct relation

Dotted line = Indirect relation

Thus the idea is that the tātparya knowledge produced in the addressee is a kind of knowledge indicated by the expression conditioned through utterability characterised by a desire (in the form of intention and understandability)

It is to be noted that the Naiyāyikas always speak in terms of valid expressions or words and accordingly they put forward their theory. As to the nature of validity Jagadīśa observes that the expressions or words which present their meaning in association with embedded constituents are termed as valid. In other words, the words or expressions which project their associative meaning in particular faculty of import are termed as valid. For instance, *Pākaḥ* √(Pac-ghañ) means 'act of cooking' which is an associative meaning. The importance of tātparya is mostly realised in case of double meaning and multimeaning expressions. For instance,

1. *Saindhavam ānaya* 'Bring the horse/salt'.

   In this case, since the term *saindhava* has two meanings the addressee has to understand the intention of the speaker as to whether he wants salt or horse.

2. *Harim paśya* 'See the Lord Hari/monkey/lion etc.'.

   In this case, the word *Hari* has a number of meanings. The addressee has to decide one meaning taking note of the intention of the speaker.

Although these cases may be disposed of through context and such other conditions, the context etc. cannot be taken as causal factor owing to their non-stable nature. The 'intention' is therefore, taken into consideration as an ancillary factor as it is less prolix and more judicious to serve the purpose. Moreover, in case of expression of non-human agency such as vedic expression, the desire of *Īśvara* 'God' is taken into account through interpretation.

Tātparya as an Expressive Power :

Jayant Bhaṭṭa recognises ***tātparya*** as an expressive power. According to him, words being recalled and referred to by a synthetic judgement constitute a sentence. The meanings of words thus being recalled and referred to by such a such a judgement constitute the meaning of a sentence. Thus a representative judgement immediately follows the presentative one.

Jayanta does not support *anvitābhidhana* and *abhihitānvaya* theories which according to him, suffer from many defects. He points out that the power of conveying the primary meaning does not point to the related meaning. An expression has a special power which may rightly be called *tātparya*. The function of this power is to reveal the meanings of words contained in an expression as being in relation with one another. It continues to work unless and until the meanings of words are presented to our consciousness. This is a common effect of powers belonging to all words and continues so long as these words do not produce an independent judgement which represents a self-contained unity of ideas. Moreover, relational meaning is not conveyed directly by primary import. But through *tātparya* an expression becomes expressive of a correlation of the whole.

Jayanta, further, concludes that we utter words with a fact in view that they would conjointly produce their main effect but not that they would merely communicate their own primary meaning. Thus, meanings, of words being mutually related constitute the meaning of an expression.

Import and tātparya :

There is a claim that *tātparya* should be included in the faculty of import as it has certain role in the process of knowledge representation. They say that it should be accepted as a *vṛtti,* a type of relation in syntactical agreement. For instance,

Gaṅgāyāṁ ghoṣaḥ 'A cowherd's station is on the Gaṅgā'.

There is incompatibility in the meaning of such expression as it is impracticable for cowherd's station to be on the river-flow. The intention of the speaker, as they claim, therefore, stands to reason to be a factor for disambiguating the expression as 'the cowherd's station is on the bank of the Gaṅgā'.

Such a claim has been rejected on the ground that *tātparya* cannot be accepted as a *vṛtti*, because *vṛtti* is that type of relation which is an outcome of meaning constituents presented by embedded words. *Tātparya* knowledge is just an extra linguistic element directly involved in the process of knowledge representation.

The Logician's concept of *tātparya* has its main root in the speaker's desire. It is regarded as neither a syntactic relation nor a semantic tool. Since it has certain role in case of double meaning and multi-meaning situation it is accepted as an extralinguistic element contributory to the procedure of knowledge representation. Moreover, Jayanta Bhaṭṭa, the noted Logician, has propounded a theory of tātparya which points to an expressive potency of an expression under the faculty of import.

The Theory of convention:

The Naiyāyikas propound the theory of convention, known as *samaya,* which according to them, adequately explains the relation that subsists between a word and its meaning.

There are two main theories namely *autpattika* or *nitya* 'natural/eternal' propounded by the Mīmāṁsakas and *sāmayika* or *sāṁketika* 'conventional' of the Naiyāyikas. The Grammarian's theory stands very close to that of the Mīmāṁsakas.

According to the Mīmāṁsakas, a word has an innate signifying power to convey a meaning. In other words, the power of expressibility is inherent in a word, in the same manner in which the power of burning is inherent in the fire. Thus there is a natural relationship between a word and its meaning.

The Naiyāyikas do not subscribe to this view. They advocate that the relation between a word and its meaning is of conventional origin. This has been created by the will of God. In some cases, the will of men also serves the purpose.

The Mīmāṁsakas argue that the relation is not created by any one. Even if it is accepted that the convention determines the relationship the question naturally arises as to how such convention is created. The convention cannot be taken as a creation of God before creation because it is not practicable to postualte a time when there is no use of word to refer to an object. Thus it has no stand. Moreover, if the convention determines the relation of a word to its meaning then it is possible to say that any word can refer to any object. The Mīmāṁsakas, thus, go on arguing to refute the views of the Naiyāyikas.

The Naiyāyikas state that the relation between a word and its meaning, as proposed by the Mīmāṁsakas, cannot be taken as eternal because eternity of such relationship cannot be proved by any means of knowledge. It is a fact that the meanings of words are learnt from the linguistic behaviour of the senior persons. This proves that there is a convention which determines the relation and it is not eternal. Further, the Naiyāyikas do not accept different conventions created by a person as regards different uses of a word by the same person. They point out that there is a norm which, in the form of will of God in case of words of unknown origin and that of men in case of words coined by human agency, determines the relation between a word and its meaning.

The theory may be presented in the following diagram :

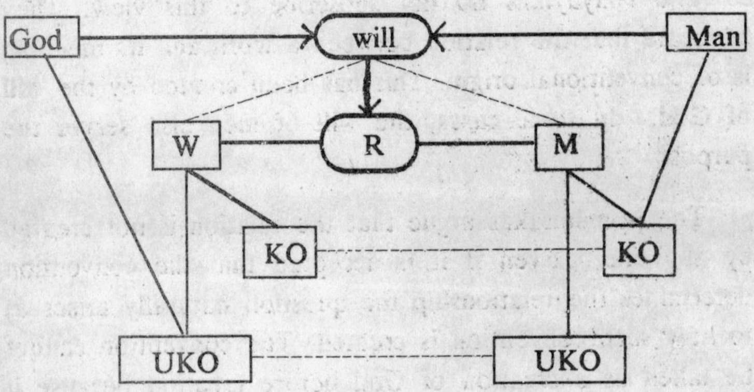

W = Word

M = Meaning.

R = Relation

KO = Known Origin

UKO = Unknown Origin

Arrow = Relational status & scope.

The main points of difference between the views of the Mīmāṁsakas and that of the Naiyāyikas may be noted in the following manner :

The Mīmāṁsakas hold that

(1) the relation between word and its meaning is eternal,

(2) the relation and the expressive power are inherent in word itself, and

(3) the relation is determined by *pratyakṣa* 'perception' *anumāna* 'inference' and *arthāpatti* 'assumption'.

The Naiyāyikas maintain that

(1) the relation between word and its meaning dates from the very dawn of creation.

(2) the relation is not inherent but exclusively grasped.

(3) two types of *pramāṇas* namely *pratyakṣa* 'perception and *anumāna* 'inference' are enough to determine the relation and

(4) the relation shapes itself into the act of will of God and man and on the strength of which the meaning of a word is grasped.

******

* पदार्थोद्देशप्रकरणम्

(1) प्रमाणतोऽर्थप्रतिपत्तौ प्रवृत्तिसामर्थ्यादर्थवत्प्रमाणम्।

(2) प्रमाणमन्तरेण नार्थप्रतिपत्तिः, नार्थप्रतिपत्तिमन्तरेण प्रवृत्तिसामर्थ्यम्। प्रमाणेन खल्वयं ज्ञाताऽर्थमुपलभ्य तमर्थमभीप्सति जिहासति वा।

(3) तस्येप्साजिहासाप्रयुक्तस्य समीहा प्रवृत्तिरित्युच्यते। सामर्थ्यं पुनरस्याः फलेनाऽभिसम्बन्धः। समीहमानस्तमर्थमभीप्सन् जिहासन् वा तमर्थमाप्नोति जहाति वा।

(4) अर्थस्तु सुखं सुखहेतुश्च, दुःखं दुःखहेतुश्च। सोऽयं प्रमाणार्थोऽपरिसङ्ख्येयः, प्राणभृद्भेदस्यापरिसङ्ख्येत्वात्। अर्थवति च प्रमाणे प्रमाता प्रमेयं प्रमितिरित्यर्थवन्ति भवन्ति।

(5) कस्मात्? अन्यतमापायेऽर्थस्यानुपपत्तेः। तत्र यस्येप्साजिहासाप्रयुक्तस्य प्रवृत्तिः स प्रमाता, स येनाऽर्थं प्रमिणोति तत्प्रमाणं, योऽर्थः प्रतीयते तत् प्रमेयं, यत् अर्थविज्ञानं सा प्रमितिः, चतसृषु चैवंविधास्वर्थतत्वं परिसमाप्यते।

(6) किं पुनस्तत्वम्। सतश्च सद्भावोऽसतश्चाऽसद्भावः। सत्सदिति गृह्यमाणं यथाभूतमविपरीतं तत्त्वं भवति। असच्चाऽसदिति गृह्यमाणं यथाभूतमविपरीतं तत्त्वं भवति।

(7) कथमुत्तरस्य प्रमाणेनोपलब्धिरिति ? सत्युपलभ्यमाने तदनुपलब्धेः प्रदीपवत्। यथा दर्शकेन दीपेन गृह्यमाणे तदिव यन्न गृह्यते, तन्नास्ति, यद्यभविष्यदिदमिव व्यज्ञास्यत,विज्ञानाभावान्नास्तीति, एवं प्रमाणेन सति गृह्यमाणे तदिव यन्न गृह्यते, तन्नास्ति, यद्यभविष्यदिदमिव व्यज्ञास्यत, विज्ञानाभावान्नास्तीति तदेवं सतः प्रकाशकं प्रमाणमसदपि प्रकाशयतीति।

---

* न्यायभाष्यत उद्धृतमिदं प्रकरणम् (न्या. भा. १/१)

(8) सच्च खलु षोडशधा व्यूढमुपदेक्ष्यते। तासां खल्वासां सद्विधानाम् – **प्रमाणप्रमेयसंशयप्रयोजनदृष्टान्तावयवतर्कनिर्णयवादजल्पवितण्डाहेत्वाभासच्छलजातिनिग्रहस्थानानां तत्त्वज्ञानान्निःश्रेयसाधिगमः।**

(9) निर्देशे यथावचनं विग्रहः। चार्थे द्वन्द्वसमासः। प्रमाणादीनां तत्त्वमिति शैषिकी षष्ठी। तत्त्वस्य ज्ञानं निःश्रेयसस्याऽधिगम इति कर्मणि षष्ठयौ। त एतावन्तो विद्यमानार्थाः, येषामविपरीतज्ञानार्थमिहोपदेशः। सोऽयमनवयवेन तन्त्रार्थ उद्दिष्टो वेदितव्यः। आत्मादेः खलु प्रमेयस्य तत्त्वज्ञानान्निःश्रेयसाधिगमः। तच्चैतदुत्तरसूत्रेणाऽनूद्यत इति।

(10) हेयं तस्य निर्वर्तकं, हानमात्यन्तिकं, तस्योपायोऽधिगन्तव्य इत्येतानि चत्वार्यर्थपदानि सम्यग्बुद्ध्वा निःश्रेयसमधिगच्छति।

(11) तत्र संशयादीनां पृथग्वचनमनर्थकम् - संशयादयो यथासम्भवं प्रमाणेषु प्रमेयेषु चान्तर्भवन्तो न व्यतिरिच्यन्त इति? सत्यमेतत्। इमास्तु चतस्रो विद्याः पृथक्प्रस्थानाः प्राणभृतामनुग्रहायोपदिश्यन्ते। यासां चतुर्थीयमान्वीक्षिकी न्यायविद्या। तस्याः पृथक्प्रस्थानाः संशयादयः पदार्थाः। तेषां पृथग्वचनमन्तरेणाऽध्यात्मविद्यामात्रमियं स्यात्, यथोपनिषदः। तस्मात् संशयादिभिः पदार्थैः पृथक् प्रस्थाप्यते। तत्र नानुपलब्धे न निर्णीतेऽर्थे न्यायः प्रवर्त्तते, किं तर्हि ? संशयितेऽर्थे। यथोक्तं - "विमृश्य पक्षप्रतिपक्षाभ्यामर्थावधारणं निर्णयः" (अ०। आ०। सू. ४१ इति। विमर्शः–संशयः। पक्षप्रतिपक्षौ–न्यायप्रवृत्तिः। अर्थावधारणं – निर्णयस्तत्त्वज्ञानमिति। स चायं किंस्विदिति वस्तुविमर्शमात्रमनवधारणं ज्ञानं संशयः प्रमेयेऽन्तर्भवन्नेवमर्थं पृथगुच्यते।

(12) अथ प्रयोजनम् - येन प्रयुक्तः प्रवर्तते, तत् प्रयोजनम्। यमर्थमभीप्सन् जिहासन् वा कर्मारभते। तेनाऽनेन सर्वे प्राणिनः सर्वाणि कर्माणि सर्वाश्च विद्या व्याप्ताः, तदाश्रयश्च न्यायः प्रवर्त्तते। कः पुनरयं न्यायः ? प्रमाणैरर्थपरीक्षणं न्यायः। प्रत्यक्षागमाश्रितमनुमानं साऽन्वीक्षा। प्रत्यक्षागमाभ्यामीक्षितस्याऽन्वीक्षणमन्वीक्षा, तया प्रवर्त्तत इत्यान्वीक्षिकी-न्यायविद्या-न्यायशास्त्रम्। यत् पुनरनुमानं प्रत्यक्षागमविरुद्धं न्यायाभासः स इति।

(13) तत्र वादजल्पौ सप्रयोजनौ। वितण्डा तु परीक्ष्यते। वितण्डया प्रवर्त्तमानो वैतण्डिकः। स प्रयोजनमनुयुक्तो यदि प्रतिपद्यते , सोऽस्य पक्षः सोऽस्य सिद्धान्त इति वैताण्डिकत्वं जहाति। अथ न प्रतिपद्यते , नायं लौकिको न परीक्षक इत्यापद्यते। अथापि परपक्षप्रतिषेधज्ञापनं प्रयोजनं ब्रवीति ? एतदपि तादृगेव यो ज्ञापयति यो जानाति येन ज्ञाप्यते यच्च ज्ञाप्यते एतच्च प्रतिपद्यते यदि ? तदा वैतण्डिकत्वं जहाति। अथ न प्रतिपद्यते? परपक्षप्रतिषेधज्ञापनं प्रयोजनमित्येतदस्य वाक्यमनर्थकं भवति। वाक्यसमूहश्च स्थापनाहीनो वितण्डा, तस्य यद्यभिधेयं प्रतिपद्यते ? सोऽस्य पक्षः स्थापनीयो भवति। अथ न प्रतिपद्यते ? प्रलापमात्रमनर्थकं भवति, वितण्डात्वं निवर्त्तत इति।

(14) अथ दृष्टान्तः प्रत्यक्षविषयोऽर्थः – यत्र लौकिकपरीक्षकाणां दर्शनं न व्याहन्यते। स च प्रमेयम्। तस्य पृथग्वचनं च तदाश्रयावनुमानागमौ-तस्मिन् सति स्यातामनुमानागमावसति च न स्याताम्। तदाश्रया च न्यायप्रवृत्तिः। दृष्टान्तविरोधेन च परपक्षप्रतिषेधो वचनीयो भवति दृष्टान्तसमाधिना च स्वपक्षः साधनीयो भवति। नास्तिकश्च दृष्टान्तमभ्युपगच्छन्नास्तिकत्वं जहाति। अनभ्युपगच्छन् किंसाधनः परमुपालभेतेति। निरुक्तेन च दृष्टान्तेन शक्यमभिधातुं– "साध्यसाधर्म्यात् तद्धर्मभावी दृष्टान्त उदाहरणम् (अ०। आ०। सू० ३६) तद्विपर्ययाद्वा विपरीतम् (अ०। आ०। सू० ३७) इति।

अस्त्ययमित्यनुज्ञायमानोऽर्थः सिद्धान्तः। स च प्रमेयम्। तस्य पृथग्वचनं–सत्सु सिद्धान्तभेदेषु वादजल्पवितण्डाः प्रवर्त्तन्ते, नातोऽन्यथेति।

(15) साधनीयार्थस्य यावति शब्दसमूहे सिद्धिः परिसमाप्यते, तस्य पंचावयवाः प्रतिज्ञादयः समूहमपेक्ष्याऽवयवा उच्यन्ते। तेषु प्रमाणसमवायः-– आगमः प्रतिज्ञा। हेतुरनुमानम्। उदाहरणं प्रत्यक्षम्। उपनयनमुपमानम्। सर्वेषामेकार्थसमवाये सामर्थ्यप्रदर्शनं निगमनमिति। सोऽयं परमो न्याय इति। एतेन वादजल्पवितण्डाः प्रवर्त्तन्ते नातोऽन्यथेति। तथाश्रया च तत्त्वव्यवस्था। ते चैतेऽवयवाः शब्दविशेषाः सन्तः प्रमेयेऽन्तर्भूता एवमर्थं पृथगुच्यन्त इति।

(16) तर्को न प्रमाणसङ्गृहीतो, न प्रमाणान्तरं, प्रमाणानामनुग्राहकस्तत्वज्ञानाय कल्पते। तस्योदाहरणम्-किमिदं जन्म कृतकेन हेतुना निर्वर्त्यते, आहोस्विद कृतकेन ? अथाकस्मिकमिति ? एवमविज्ञातेऽर्थे कारणोपपत्त्या ऊहः प्रवर्त्तते। यदि कृतकेन हेतुना निर्वर्त्यते ? हेतूच्छेदस्याऽशक्यत्वादनुपपन्नोऽयं जन्मोच्छेदः। अथाऽकृतकेन हेतुना ? ततो हेतूच्छेदस्याऽशक्यत्वादनुपपन्नो जन्मोच्छेदः। अथाऽकस्मिकम् ? अतोऽकस्मान्निर्वर्त्यमानं न पुनर्निर्वत्स्यतीति निवृत्तिकारणं नोपपद्यते, तेन जन्मानुच्छेद इति। एतस्मिंस्तर्कविषये कर्मनिमित्तं जन्मेति प्रमाणानि प्रवर्तमानानि तर्कणाऽनुगृह्यन्ते, तत्त्वज्ञानविषयस्य विभागात् तत्वज्ञानाय कल्पते तर्क इति। सोऽयमित्थम्भूतस्तर्कः प्रमाणसहितो वादे साधनायोपालम्भाय चाऽर्थस्य भवतीत्येवमर्थपृथगुच्यते प्रमेयान्तर्भूतोऽपीति।

(17) निर्णयस्तत्त्वज्ञानं प्रमाणानां फलं, तदवसानो वादः। तस्य पालनार्थे जल्पवितण्डे। तावेतौ तर्कनिर्णयौ लोकयात्रां वहत इति। सोऽयं निर्णयः प्रमेयान्तर्भूत एवमर्थं पृथगुदिष्ट इति।

(18) वादः खलु नार्नाप्रवक्तृकः प्रत्यधिकरणसाधनोऽन्यतरा धिकरणनिर्णयावसानो वाक्यसमूहः पृथगुद्दिष्ट उपलक्षणार्थम्। उपलक्षितेन व्यवहारस्तत्त्वज्ञानाय भवतीति। तद्विशेषौ जल्पवितण्डे तत्त्वाध्यवसायसंरक्षणार्थमित्युक्तम् (अ० ४आ०२ सू०५०)

(19) निग्रहस्थानेभ्यः पृथगुद्दिष्टा हेत्वाभासा वादे चोदनीया भविष्यन्तीति। जल्पविण्डयोस्तु निग्रहस्थनानीति। छलजातिनिग्रहस्थानानां पृथगुपदेश उपलक्षणार्थमिति। उपलक्षितानां स्ववाक्ये परिवर्जनं छलजातिनिग्रहस्थानानां परवाक्ये पर्यूर्यनुयोगः। जातेश्च परेण प्रयुज्यमानायाः सुलभः समाधिः, स्वयं च सुकरः प्रयोग इति।

(20.) सेयमान्वीक्षिकी प्रमाणादिभिः पदार्थैर्विभज्यमाना –

प्रदीपः सर्वविद्यानामुपायः सर्वकर्मणाम्।
आश्रयः सर्वधर्माणां विद्योद्देशे प्रकीर्त्तिता।।

तदिदं तत्त्वज्ञानं निःश्रेयाधिगमश्च यथाविद्यं वेदितव्यम्। इह त्वध्यात्मविद्यायामात्मादितत्त्वज्ञानं तत्त्वज्ञानम्।

निःश्रेयसाधिगमोऽपवर्गप्राप्तिरिति ।।-१- ।।

* ENUMERATION SECTION

1. The object is cognised by *pramāṇa* 'the means of valid knowledge', which leads to successful activity. It is for this reason that the validity of *pramāṇa* is ascertained in point of an invariable relation realised by the object it points to.

2. There is no cognition without the means of valid knowledge. There is no successful activity without the cognition of object. It is only with the help of *pramāṇa* the knower, being cognizant of the object, desires or discards the same.

3. His effort prompted by the desire of obtainment and relinquishment is called activity the success of which is related to the result. The endeavourer being desirous of obtainment or relinquishment obtains or discards it.

4. The object means pleasure and its cause as well as suffering and its cause. The object of *pramaṇa* is innumerable because the distinction of living beings is uncountable. When the pramāṇa is validated the validity of knower, knowable and knowledge is established in relation to the object.

5. How it is so ? Because, in the absence of the foremost, namely *pramāṇa* the true knowledge can not be effected. In this case, he is the *pramatā* 'knower' who possesses a desire to seek or shun the object. The *pramāṇa*

* **This Chapter has been taken from Nyāya-bhāṣya (1.1)**

is the instrument by which the knower rightly knows the object. That which is presented to consciousness is the *prameya* 'object'. The *pramiti* is the right knowledge of the object. The science of knowledge and its content gets fulfilment with these four.

6. What is then this science ? It refers to positivity of positive and negativity of negative. When the positive object is known as positive without any contradiction, it becomes itself the science. Similarly, when the negative object is known as negative without any contradiction, it becomes the science.

7. But how can the latter, namely the nagative, is ascertained by the *pramāṇa ?* It is just like the lamp with the help of which, the object which exists, is known so also the object which does not exist,is known as not existing. Just as with the help of the lamp the visible object is ascertained by the observer, similarly the object which is not perceived by the same means becomes non-existent. Had it been there like the visible object it would have been perceived. Since there is no perception (of it) it is ascertained as non-existent. Thus, when a positive object is known by a pramāṇa, that which at the same time is not ascertained is known as non-existent. If it were there, it would have been known just as the positive object. It is ascertained as non-existent because it is not perceived. Thus, the pramāṇa which reveals the existent also reveals the non-existent.

8. The existents would be stated briefly in sixteen forms. The true knowledge of *pramāṇa* (valid means of knowledge) *prameya* (object) *saṁśaya* (doubt) *prayojana* (incentive) *dṛṣṭānta* (corroborative instance) *siddhānta* (proved

doctrine) nirṇaya (final ascertooainment) vāda (discussion for the final ascerrtnment) *jalpa* (debating maneuver) *vitaṇḍā* (destructive criticism) *hetvābhāsa* (pseudoprobans) *chala* (Purposive distortion of the opponent) *jāti* (futile rejoinder based on similarity or dissimilarity) and *nigrahasthāna* (point of defeat) leads to the attainment of the highest good.

9. The underlying structure of this *sūtra* is to be construed in the following way : The categories are to be put in *dvandva* compound. *Pramāṇādīnām tattvam* refers to genitive relation. *Tattvasya jnanam,* and *niḥśreyasasya adhigamaḥ* refer to objectivized - genitive relation. These are the sixteen positive objects which aim at non-contradictory knowledge. It is to be understood that the objects to be discussed in Nyāya Philosophy are exhaustively enumerated each by name. The highest good is achieved through the true knowledge of the objects like *ātman* etc. This has been explained in the next *sūtra.*

10. One attains the highest good after understanding the four basic human concerns :

(i) removable (ie.e suffering) and its causes.

(ii) the cause of the absolute cessation of suffering.

(iii) the means of attaining the right knowledge.

(iv) the (ultimate) acheivable goal.

11. The opponent argues : The separate mention of 'doubt' etc. is useless as included under **pramāṇa** and *prameya* etc. They do not show any distinction. The answer is : This is true. The four branches of study, each having its unique character are prescribed for the benefit of mankind. Amongst these the study of logic is the fourth

and its unique subject matter (*prasthāna*) consists of these categories namely 'doubt' etc. Without mention of these 'logic' would have been mere the 'study of the self' like Upaniṣadas. It is for this reason that the categories like 'doubt' etc. have been separately mentioned.

The structure of logic does not function in the cases that are unknown, nor has its relevance for those that are certainly known. How does its relevance stand ? It has relevance only for those objects about which there is doubt. This is stated in Nyāya Sūtra (1.1.41) that final ascertainment is the ascertainment of an object through consideration of thesis and antithesis which result from doubt. Here *vimarśa* means doubt. Thesis and antithesis mean application of the structure of Logic. Ascertainment of the object means truc knowledge. The form of doubt is "what exactly is it ?"

This knowledge pertains to all concerned objects in a form of non-ascertainment. Although included in *prameya* category it is separately mentioned as being the basic condition of application of the structure of logic.

12. Next is the incentive (*Prayojana*). What induces one to activity is the incentive. Or it is the object which leads to activity through a desire of attainment or relinquishment. All living beings, all their activities, and all branches of learning are pervaded by incentive. The structure of logic also function taking recourse to its very basis. What is then this *Nyāya* 'the sturcture of Logic' ? *Nyāya* is the examination of an object with the help of valid means of knowledge. The inference, which resorts to perception and scripture is called *anvīkṣā*. That which is already known by perception and scriptural attestation is

anvikṣā (knowing over again). The branch of knowledge led by this *anvīkṣikī* is called Nyāya vidyā or Nyāya-śāstra. The inference which is contradicted by either perception or scripture is pseudo-structure of Logic.

13. 'Discussion for the final ascertainment' and 'Debating maneuver' are useful there. But the purposefulness of destructive criticism is being examined. A destructive critic is described as one who employs destructive criticism. If he goes on admiting his purpose, such and such to be his thesis and such and such to be his doctrine he surrenders his status of being a destructive critic. If he does not admit he becomes neither a lay learner nor an expert. If he says that his purpose is to expose the faults of the apponent, this would lead to renounce his position by admitting 'one who expounds, one who learns, the means of expounding or the theme expounded., If he does not admit his contention that his purpose is to refute the position of the opponent becomes futile.

Vitaṇḍā means certain propositions without any positive thesis. If these propositions are claimed to be meaningful then this constitute his positive thesis. If this is not admitted the propositions become meaningless like delirium and as such the essence of *Vitandā* is lost.

14. Corroborative instance is an object of perception in which the realisation of both lay learner and the expert does not stand in conflict. This is included in the *prameya* category. It is separately mentioned because it is the basis of inference and verbal testimony. If it is there then both inference and verbal testimony are possible. These are not possible without these. It also take recourse to the opplication of Nyāya. By the contradiction of *dṛṣtānta* the

position of the opponen is refuted. By substantiation of the same one's own position is established. If the *nāstika* (sceptic) admits a corroborative istance, he surrenders his scepticism. If he does not admit any how can he make his opponent silent ? By previously mentioning *dṛṣṭānta* in the present sūtra Gautama is justified in by saying subsequently : "*Udāharaṇa* is a corroborative instance which being similar to the subject (*sādhya-dharmin*) possessed its characteristics (*tat-dharma-bhāvin*) 1.1.36." and 'because of its absence the opposite of that 1.1.37".

Siddhānta (proved doctrine) means the admitted in the form "It exists". It is included in the *prameya* category. It is separately mentioned because *vāda, jalpa* and *vitaṇḍā* are employed only when there is difference in the proved doctrines, not otherwise.

15. The five propositions (namely *pratijñā* and *hetu* etc.) with which the desired thesis is conclusively established are called components of inference 'in relation to their totality., The *pramāṇas* inhere in these five parts. The preliminary proposition is the verbal testimony. The proban is inference. The examplification is perception. The application is comparison.

The demonstration of capability of all these components for establishing the main thesis is the conclusion. Such is the *Nyāya par excellence.* With help of this *vāda, jalpa* and *vitaṇḍā* can be employed, never otherwise. The ascertainment on truth resorts to it. These inference components are specific form of propositions and as such are included in *prameya* category. They are mentioned separately for the above reasons.

16. The hypothetical argument is not included in

Pramāṇas. It is also not a separate *pramāṇa.* It is just an accessory to the valid means of knowledge contributing to the ascertainment of truth. The example is "Is this birth produced by an impermanent cause or by a permanent one or accidental ?" In this case in which the real nature of the object is unknown *ūha* 'approximative argument' is employed by way of adducing some reasons - " If birth is produced by an impermanent cause, the cessation of birth becomes impossible when the cause is destroyed. Again, if it is produced by a permanent cause, then, because of impossibility of the destruction of cause, the cessation of birth-series becomes impossible. Again, if it is accidental, then birth, being accidentally produced will never cease. Since there is no cause of cessation there will be no cessation of birth. " In such a case of *tarka,* the *pramāṇas* which prove that birth is due to ***karman*** are favoured by a *tarka.* Since *tarka* judges the plausibility of the contentent of the knowledge of truth it facilitates the true knowledge. Even if tarka is included in *prameya* category it is mentioned separately because it, alongwith the *pramāṇas* helps to establish, one's own thesis and to refute that of the opponent in a debate.

17. 'Final ascertainment' is the knowledge of truth. This is the result of valid means of knowledge. The *vāda* 'discussion for the final ascertainment' ends with this.

Jalpa and *vitaṇḍā* are intened to nourish it. The *tarka* and *nirṇaya* conduct the practical life of the people. It is for this reason only that the *nirṇaya* is separately mentioned although it is included in *prameya* category.

18. 'Discussion for final ascertainment' means the arguments offered by more than one party - each trying to

establish its own thesis - ending in the ascertainment of one of the positions contended. It is mentioned separately because of the need of its special knowledge. When the *vada* is specially known its employment yields the knowledge of truth. *Jalpa* and *Vitaṇḍā* wich differ from it in certain respects are mentioned separately because these are for the protection of the knowledge of truth as said by Gautam in the *Sūtra* 4.2.50.

19. The 'pseudo probans' are mentioned over and above the 'point of defeat' because these are to be employed in *vāda* while *nigrahasthānas* are to be employed in *jalpa* and *vitaṇḍā*. The separate mention of *chala, jāti* and *nigrahasthāna* is for the shake of acquintance with these. *Chala,* jāti and *nigrahasthāna* when acquainted with, can be avoided in one's own argument and detected in those of others. Further, it becomes easier to answer a *jāti* employed by the apponent and one may easily use it for one's own purpose.

20. The *anvīkṭikí* which is categorised by the *pramaṇas* etc. is specially mentioned in the enumeration of various studies as the lamp for all branches of learning, the means for all activities and the basis of all virtues.

The 'knowledge of truth' and the 'attainment of highest good' are to be construed in accordance with the respective branches of learning. In this particular branch of learing relating to self the 'knowledge of truth' means the wisdom of self and the 'attainment of highest good' means 'liberation'.

* शब्दशक्तिपरीक्षाप्रकरणम्

(1) ते विभक्त्यन्ताः पदम् ।।५७।।

यथादर्शनं विकृता वर्णा विभक्त्यन्ताः पदसंज्ञा भवन्ति। विभक्तिर्द्वयी नामिक्याख्यातिकी च, ब्राह्मणः पचतीत्युदाहरणम्। उपसर्गनिपातास्तर्हि न पदसंज्ञा लक्षणान्तरं वाच्यम् इति? शिष्यते च खलु नामिक्या विभक्तेरव्ययाल्लोपः तयोः पदसंज्ञार्थमिति ।।-५७-।।

(2) पदेनार्थसम्प्रत्यय इति प्रयोजनम्, नामपदं चाधिकृत्य परीक्षा, गौरिति पदं खल्विदमुदाहरणम् –

तदर्थे व्यक्त्याकृतिजातिसन्निधावुपचारात्संशयः। ।।-५८-।।

अविनाभाववृत्तिः सन्निधिरिति। अविनाभावेन वर्तमानासु व्यक्त्याकृतिजातिषु गौरिति प्रयुज्यते, तत्र न ज्ञायते किमन्यतमः पदार्थ उत सर्व ? इति ।।-५८।।

(3) शब्दस्य प्रयोगसामर्थ्यात् पदार्थावधारणं तस्मात –

याशब्दसमूहत्यागपरिग्रहसङ्ख्यावृद्ध्यपचय-वर्णसमासानुबन्धानां व्यक्तावुपचाराद् व्यक्तिः ।।-५९-।।

व्यक्तिः पदार्थः, कस्मात् ? याशब्दप्रभृतीनां व्यक्तावुपचारात्। उपचारः प्रयोगः।

या गौस्तिष्ठति या गौर्निषण्णेति, नेदं वाक्यं जातेरभिधायकमभेदात्। भेदात्तु द्रव्याभिधायकम्। गवां समूह इति भेदाद् द्रव्याभिधानम्, न जातेरभेदात्।

चैद्याय गां ददातीति द्रव्यस्य त्यागो न जातेरमूर्त्तत्वात्, प्रतिक्रमानुक्रमानुपपत्तेश्च।

परिग्रहः -स्वत्वेनाऽभिसम्बन्धः कौण्डिन्यस्य गौर्ब्राह्मणस्य गौरिति, द्रव्याभिधाने द्रव्यभेदात् सम्बन्धभेद इति उपपन्नम्, अभिन्ना तु जातिरिति।

---

* न्यायभाष्यत उद्धृतमिदं प्रकरणम् (न्या. भा. 2/2)

सङ्ख्या, दश गावो विंशतिर्गाव इति भिन्नं द्रव्यं सङ्ख्यायते, न जातिरभेदादिति।

वृद्धिः- कारणवतो द्रव्यस्यावयवोपचयः, अवर्द्धत गौरिति। निरवयवा तु जातिरिति। एतेनापचयो व्याख्यातः।

वर्ण :- शुक्ला गौः कपिला गौरिति, द्रव्यस्य गुणयोगो न सामान्यस्य ।

समास :- गोहितं गोसुखमिति द्रव्यस्य सुखादियोगो न जातेरिति।

अनुबन्ध :- **सरूपप्रजनसन्तानो** गौर्गां जनयतीति, तदुत्पत्तिधर्मत्वाद् द्रव्ये युक्तं न जातौ विपर्ययादिति। द्रव्यं व्यक्तिरिति हि नार्थान्तरम्।।-५९-।।

(4) अस्य प्रतिषेधः -

न तदनवस्थानात्।।-६०-।।

न व्यक्तिः पदार्थः। कस्मात्? अनवस्थानात्। या शब्दप्रभृतिभिर्यो विशेष्यते स गोशब्दार्थो या गौस्तिष्ठति या गौनिषण्णेति, च द्रव्यमात्रमविशिटं जात्या विनाऽभिधीयते। किं तर्हि ? जातिविशिष्टम्। तस्मान्न व्यक्तिः पदार्थः। एवं समूहादिषु द्रष्टव्यम्।।६०।।

(5) यदि व्यक्तिः न पदार्थः कथं तर्हि व्यक्तावुपचार इति ? निमित्तादतद्भावेऽपि तदुपाचारः। दृश्यते खलु –

सहचरणस्थानतादर्थ्यवृत्तमानधारणसामीप्ययोगसाधनाधिपत्येभ्यो ब्राह्मणमञ्चकटराजसक्तुचन्दनगङ्गाशाटकान्नपुरुषेष्वतद्भावेऽपि। तदुपचारः-६१-।।

अतद्भावेऽपि तदुपचार इति - अतच्छब्दस्य तेन शब्देनाभिधानमिति।

सहचरणात् - यष्टिकां भोजयेति, यष्टिकासहचरितो ब्रह्मणोऽभिधीयत इति।

स्थानात् - मंचाः क्रोशन्तीति मंचस्थाः पुरुषा अभिधीयन्ते।

तादर्थ्यात् - कटार्थेषु वीरणेषु व्यूह्यमानेषु कटं करोतीति भवति।

वृत्ताद् - यमो राजा कुवेरो राजेति तद्वद्वर्त्तते इति।

मानाद् - आढकेन मिताः सक्तवः आढकसक्तव इति।

धारणात् - तुलायां धृतं चन्दनं तुलाचन्दनमिति।

सामीप्याद् - गंगायां गावश्चरन्तीति देशोऽभिधीयते सन्निकृष्टः।

योगात् - कृष्णेन रागेण युक्तः शाटकः कृष्ण इत्यभिधीयते।

साधनात् - अन्नं प्राणा इति।

आधिपत्यात् - अयं पुरुषः कुलम् अयं गोत्रमिति। तत्रायं सहचरणाद्योगाद् वा जातिशब्दो व्यक्तौ प्रयुज्यत इति।।-६१।।

(6) यदि गौरित्यस्य पदस्य न व्यक्तिरर्थः, अस्तु तर्हि -

आकृतिस्तदपेक्षत्वात् सत्त्वव्यवस्थानसिद्धेः ।।.६२।।

आकृतिः पदार्थः। कस्मात् ? तदपेक्षत्वात् सत्त्ववव्यवस्थानसिद्धेः। सत्वावयवानां तदवयवानां च नियतो व्यूह आकृतिः तस्यां गृह्यमाणायां सत्त्वव्यवस्थानं सिद्ध्यति अयं गौरयमश्व इति, नागृह्यमाणायाम्। यस्य ग्रहणात् सत्त्वव्यवस्थानं सिद्ध्यति तं शब्दोऽभिधातुमर्हति सोऽस्यार्थ इति।

नैतदुपपद्यते यस्य जात्या योगस्तत्र जातिविशिष्टमभिधीयते गौरिति। न चावयवव्यूहस्य जात्या योगः। कस्य तर्हि? नियतावयवव्यूहस्य द्रव्यस्य। तस्मान्नाकृतिः पदार्थः ।।.६२.।।

(7) अस्तु तर्हि जातिः पदार्थः-

व्यक्त्याकृतियुक्तेऽप्यप्रसङ्.गात् प्रोक्षणादीनां मृद्गवके जातिः।।-६३-।।

जातिः पदार्थः। कस्मात्? व्यक्त्याकृतियुक्तेऽपि मृद्गवके प्रोक्षणादीनामप्रसङ्.गादिति। गां प्रोक्षय गामानय गां देहीति नैतानि मृद्गवके प्रयुज्यन्ते। कस्मात् ? जातेरभावात्। अस्ति हि तत्र व्यक्तिः, अस्त्याकृतिः यदभावात्तत्रासम्प्रत्ययः स पदार्थ इति।।-६३-।।

(8) **नाकृतिव्यक्त्यपेक्षत्वाज्जात्यभिव्यक्तेः।।-६४-।।**

जातेरभिव्यक्तिराकृतिव्यक्ती अपेक्षते, नागृह्यमाणायामाकृतौ व्यक्तौ जातिमात्रं शुद्धं गृह्यते, तस्मान्न जातिः पदार्थ इति ।।।-६४-।।

(9) न वै पदार्थेन न भवितुं शक्यं, कः खल्विदानीं पदार्थ इति?
व्यक्त्याकृतिजातयस्तु पदार्थः।।-६५-।।

तुशब्दो विशेषणार्थः। किं विशिष्यते? प्रधानाङ्गभावस्यानियमेन पदार्थत्वमिति। यदा हि भेदविवक्षा विशेषगतिश्च, तदा व्यक्तिः प्रधानमङ्गं तु जात्याकृती। यदा तु भेदोऽविवक्षितः सामान्यगतिश्च, तदा जातिः प्रधानमङ्गं तु व्यक्त्याकृती। तदेतद्बहुलं प्रयोगेषु। आकृतेस्तु प्रधानभाव उत्प्रेक्षितव्यः।।-६५-।।

(10)कथं पुनर्ज्ञायते नाना व्यक्त्याकृतिजातय इति? लक्षणभेदात्। तत्र तावत् –

व्यक्तिर्गुणविशेषाश्रयो मूर्तिः।।-६६-।।
व्यज्यत इति व्यक्तिरिन्द्रियग्राह्येति न सर्वं द्रव्यं व्यक्तिः।

यो गुणविशेषाणां स्पर्शन्तानां गुरुत्वघनत्वद्रवत्वसंस्काराणामव्यापिनः परिमाणस्याश्रयो यथासम्भवं तद् द्रव्यं मूर्तिः, मूर्च्छितावयवत्वादिति ।।-६६-।।

(11)आकृतिर्जातिलिङ्गाख्या।।-६७-।।

यया जातिर्जातिलिङ्गानि च प्रख्यायन्ते तामाकृतिं विद्यात्। सा च नान्या सत्त्वावयवानां तदवयवानां च नियताद् व्यूहादिति। नियतावयवव्यूहाः खलु सवत्त्वायवा जातिलिङ्गम्, शिरसा पादेन गामनुमिन्वन्ति। नियते च सत्त्वावयवानां व्यूहे सति गोत्वं प्रख्यायत इति। अनाकृतिव्यङ्ग्यायां जातौ मृत्सुवर्णं रजतम् इत्येवमादिष्वाकृतिर्निवर्तते जहाति पदार्थत्वमिति ।।-६७-।।

(12)समानप्रसवात्मिका जातिः।।-६८-।।

या समानां बुद्धिं प्रसूते भिन्नेष्वधिकरणेषु यया वहूनीतरेतरतो न व्यावर्तन्ते योऽर्थोऽनेकत्र प्रत्ययानुवृत्तिनिमित्तं तत्सामान्यम्। यच्च केषांचिदभेदं कुतश्चिद्भेदं करोतीति तत् सामान्यविशेषो जातिरिति।।-६८-।।

* MEANING SECTION

1. "They (the groups of letters) ending with suffixes are termed as words (2.2.57)".

The group of letters ending with suffixes, undergoing transformations according to rules are termed as words. Suffixes are of two types: (1) Nominal and (2) Verbal. For instance, (1) *brāhmaṇaḥ (brāhmaṇa-su)* and (2) *pacati (√pac-ti)*, "The brahmin cooks".

Objection : Is it that since prefixes and particles are not included under the definition of word, another definition should be made ?

Answer : The elision of the suffixes in case of the indeclinable and particles is prescribed in the grammar for extending the definition of word to include indiclinable and particle.

2. The purpose is the determination of meaning from a word. The critical examination centres around a nominal term. The term cow is to be taken as the instance.

While ascertaining the meaning of a word "There is doubt as the meaning of a word is invariably related to the individual, the shape and the universal (2.2.58)".

The word *sannidhi* means invariable relation. The word cow is used to refer to the individual the shape and the universal which are invariably related to it. It is however,

---

* This Chapter has been taken from Nyāya-bhāṣya (2/2)

not known whether the meaning (of the word) is any one of these three or all the three.

3. The meaning of a word is determined from its usability. Therefore - (objection) :

"The word means individual because the word is used to refer to individual in following cases such as :

by the pronoun 'that'

by group

by gift

by acceptance of gift

by association of number

by growing

by decaying

by association of colour

by the use of compound, and

by procreation (2.2.59)"

The word refers to individual. Why ? Because the pronoun 'that' is used to refer to individual. The 'use' is signified as 'referring'.

The expression 'The cow that stands' or (The cow that is sitting' does not point to universal because there is no difference in universal. But when there is a distinction this refers to individual (cow).

In the expression 'group of cows' the word cow refers to the individual because these are different but not to universal because it is the same.

In the expression 'A person gives a cow to the learned' the thing relinquished does not refer to universal because the universal is abstract and can not be the object of ceremony preceding and following the gift.

Acceptance means making something one's own. In the expression 'The cow belongs to Kauṇḍinya' or The cow belongs to the Brahmin" the word cow refers to individual because the difference in relation due to difference of individuals is established. But the universal stands in terms of non-difference.

In the expressions 'ten cows' and 'twenty cows' the number is counted in association with the individuals, but not with the universal which has no differences.

'Growing' means the nourishment of limbs of an object having parts. The word cow, in the expression 'The cow has grown' does not refer to universal since it has no parts.

This explains also the case of decay.

In case of colour, the expression - ' a white cow' and 'a brown cow' points to the quality which can only be associated with individual (substance) but not with universal.

when there is compound like 'the welfare of the cow' and 'the pleasure of the cow', welfare etc. can be related to the individual and not to the universal.

'Procreation' means possessing the characteristic of being born and giving birth to offsprings of same category. For instance, 'a cow is born of cow'. Procreation can refer to the individuals which have the characteristic of being born and not to the universal, because it does not possess such characteristic.

The term *dravya* points to nothing but the individual.

4. This is refuted (Answer) :

"No, because of the absence of any fixed norm of the individual (2.2.60)".

The word does not refer to mere individual. Why ? Beause of the absence of any norm of the individual. That which is referred to by the pronoun 'that' etc. constitutes the meaning of the word cow. In the exprssions 'the cow that stands' and 'the cow that sits' the word 'cow' does not refer to bare individual without being characterised by the universal. therefore, mere individual is not the meaning of a word.

The cases of 'group' etc. are to be explained thus.

5. If the individual is not the meaning, then how is it referred to by a word ? In spite of the basence of the capacity to refer to the individual the words are used to refer individuals because of some specific conditions.

It is observed that

"In spite of absence of the capacity to refer, the words are used to refer to a brahmin, the platform, straw-mat, the king, the enemy, the sandal wood, the Ganges, a gown, food and a person, on the conditions such as - association, place, need, behaviour, measurment, container, nearness, connection, instrument and prominence (2.2.61)"

The expresssion *atad-bhāve'pi tadupacāra* means that something being not the meaning is referred to by the word. for instance,

Due to association : In the expression 'Feed the sticks', the sticks refer to the brahmin carrying the sticks.

Due to place : In the expression "The platforms are crying", the platform refers to the persons on the platform.

Due to need : When the strans are woven for the purpose of making a mat the expression used is, 'One is weaving a mat'.

Due to behaviour : The expressions, 'The king is a *Yama'*, or The king is a *Kuvera'* refer to behaviour.

Due to measurement : The expression 'An *āḍhaka* of pounded barley 'refers to that amount of pounded barley as measured by an *āḍhaka*.

Due to the container : The expression , 'sandal wood in the balance' refers to sandal wood intended to be weighed by the balance.

Due to nearness : In the expression "The cows graze in the Ganges' the word *Gaṅgā* refers to the field on the bank of the Ganges.

Due to connection : The expression 'Black (cloth)' refers to a cloth with black colour.

Due to instrument : In the expression 'Food is life' the food is considered as an instrument for subsistance.

Due to prominence : There are expressions, like 'This person is family' and 'This person is *gotra*'.

The word cow referring to universal is secondarily used to refer the individual because of two conditions such as : association and connection.

6. If the word 'cow' does not mean the individual then let it mean - (objection) :

"The shape, because the knowledge of existence of the animal depends on it (2.2.62)"

The word refers to the shape. Why ? Because the knowledge of the existence of the animal is dependent on it. Shape means the structure constituted by the parts of the animal and particles of those parts. When this shape is known there is definite knowledge of the existence of the animal as 'this is a cow' or 'this is a horse'. When the shape is not known there is no such cognition. That, the knowledge of which leads to the definite knowledge of the existence of the animal as 'this is a cow' or 'this is a horse'. When the shape is not known there is no such cognition. That, the knowledge of which leads to the difinite knowledge of the existence of the animal constituted the meaning referred to by a word. The shape therefore, is referred tb by the word cow.

This is not tenable. The word cow refers to the object characterised by the universal. The universal has no connection with the structure of limbs. With which then it is connected ? It is connected with the animal as a whole, constituted by the structure of limbs. The word, therefore, does not refer to the shape.

7. Then, let the word refer to the universal - (Objection) :

"The word refers to the universal, because inspite of the presence of individuality and shape in an earthen cow it is not used as the object of the ceremony of sprinkling water etc. (2.2.63)".

The word refers to the individual. Why ? Because in spite of the presence of the individual and shape in an earthen cow, it is not used as the object of the ceremony of sprinkling water etc. The expressions 'Sprinkle water on the cow', 'Bring the cow' and 'Give away the cow' are not used with reference to an earthen cow. Why ? .Because, there is no universal in it. Both individuality and shape exist in it. That in the absence of which there is no right knowledge, constitutes the meaning of the word cow.

8. (Answer):

"No, the knowledge of the universal is dependent on the shape and the individual (2.2.64)".

The knowledge of the universal depends on the shape and the individual. Without comprehension of the shape and the individual the bare universal cannot be comprehended. The word, therefore, does not refer mere to the universal.

9. It is not that the word does not refer to anything. What, then, is the referent ?

"The individual, the shape and the universal constitute the meaning of a word (2.2.65)".

The term *tu* in the *sūtra* is used to convey 'qualificative significance.' What is qualificatively signified? The prominence and subservience of either of these three as there is no fixed rule to this effect. When, by the word, the difference is sought to be expressed and it is known as a particular member, the individual stands prominent and the universal and the shape subserve relatively as subordinates. When, however, the difference is not shought to be expressed and the class is known as a whole, the universal stands prominent and the individual and the shape subserve relatively as subordinates. There are many examples of. this type. The prominence of the shape may be understood accordingly.

10. How, then it is to be ascertained that the individual the shape and the universal are different ? From the differences in their difinitions. There is -

"Individual means the specific image which is the substratum of some qualities (2.2.66)".

That which is manifested, i.e., becomes the object of the sense organ is the individual. All substances are not individual. Individual means that substance which is the substratum of qualities beginning with colour and ending with touch as well as heaviness, density, impression and non-pervasive magnitude. Because the specific image is constituted by parts conjoined together.

11. "Shape is that by which the mark of universal is known (2.2.67)".

That, by which the universal and the marks of the universal are indicated, should be known as the shape. This shape is nothing but the fixed structure constituted by the limbs of the animal and the parts of those limbs. The limbs of the animal, which are themselves constituted by a fixed combination of parts, are the marks for inferring the universal. For instance, people infer the cow from its head or feet. If there is a fixed structure of the limbs of the cow the universal cowness is manifested. Where the universal is not revealed by the shape as in the case of clay, gold and silver etc. the shape is excluded and this does not form of the part of the meaning.

12. "The universal is that which produces a cognition of commonness (2.2.68)."

That which produces the cognition of commonness in different objects, that by the presence of which different objects are not mutually differentiated, and that entity which is the cause of the continuation of the same knowledge in different objects, is the universal. That which points to identity of something with some other individuals and at the same time differentiates it from other individuals is also a universal of special type.

* शब्दार्थसम्बन्धनिरूपणम्

यदपि सम्बन्धकरणे पुरुषानपेक्षत्वमुच्यते चित्रभानोरिव दहनशक्तिः शब्दस्य नैसर्गिकी वाचकशक्तिः। व्युत्पत्तिस्तु वृद्धेभ्य एव व्यवहरमाणेभ्य उपलभ्यते इति किमत्र पुरुषः करिष्यतीति तदप्यघटमानम्, पुरुषपरिघटितसमयसम्बन्धव्यतिरेकेण शब्दादर्थप्रत्ययानुत्पत्तेः।

ननु नैव शब्दस्यार्थेन सम्बन्धः कश्चिदस्ति कस्येदं पुरुषसापेक्षत्वं तन्निरपेक्षत्वं वा चिन्त्यते, न हि शब्दार्थयोः कुण्डबदरयोरिव संयोगत्वभावः तन्तुपटयोरिव समवायात्मा वा सम्बन्धः प्रत्यक्षमुपलभ्यते, तन्मूलत्वाच्च सम्बन्धान्तराण्यपि न सन्ति, तदुक्तं मुखे शब्दमुपलभामहे भूमावर्थमिति, नाप्यनुमीयते शब्दस्यार्थेन सम्बन्धः क्षुरमोदकशब्दोच्चारणे मुखस्य पाटनपूरणानुपलम्भात् "न च शब्ददेशे अर्थः सम्भवति न चार्थदेशे शब्दः स्थानकरणप्रयत्नानां तद्धेतूनां घटाद्यर्थदेशेऽनुपलम्भात्, व्यापकत्वं तु शब्दस्य प्रतिषिद्धमेव।

उच्यते, न संश्लेषलक्षणः शब्दार्थसम्बन्धो ऽस्माभिरभ्युपगम्यते, तत् किं कार्यकारणनिमित्तनैमित्तिकाश्रयाश्रयिभावादयः श्ब्दस्यार्थेन सम्बन्धाः, एतेऽपि नतराम्, न तर्हि तस्य कश्चिदर्थेन सम्बन्धः, न नास्ति शब्दस्यार्थेन सम्बन्धः प्रत्ययनियमहेतुत्वाद्, धूमादिवत्, तत् किं शब्दार्थयोरविनाभावः सम्बन्धः, सोऽपि नास्ति, एवं हि शब्दोऽनुमानमेव स्यात्, कस्तर्हि समय इति ब्रूमः, कोऽयं समयो नाम, अभिधानाभिधेयनियमनियोगः समय उच्यते, यद्येवं किमनाशङ्कनीयसंश्लेषपरिचोदनेन तद्दूषणेन च, उच्यते, शब्दार्थाभेदवादिनां हि वैयाकरणानामेष संश्लेष उपपत्तिमान् समयोऽप्ययमनुपपन्न एव, स हि पुरुषकृतः सङ्केतो न च पुरुषेच्छया वस्तुनियमोऽवकल्पते तदच्छिाया अव्याहतप्रसरत्वात्, अर्थोऽपि किमिति वाचको न भवति, न चैवमस्ति न हि दहनमनिच्छन्नपि पुरुषो धूमान्न तं प्रत्येति जलं वा तत इच्छन्नपि प्रतिपद्यते तत्र यथा धूमाग्न्योर्नैसर्गिक एवाविनाभावो नाम सम्बन्धः ज्ञप्तये तु भूयोदर्शनादिनिमित्तमाश्रीयते एवं

* न्यायामञ्जरीत उद्धृतमिंद प्रकरणम् (न्या. म.४)

शब्दार्थयोस्सांसिद्धिक एव शक्त्यात्मा सम्बन्धः तद्व्युत्पत्तये तु वृद्धव्यवहारप्रसिद्धिसमाश्रयणम्, स्वाभाविके सम्बन्धे सति दीपादिवत् किं तद्व्युत्पत्त्यपेक्षणेनेति चेद् न शब्दस्य ज्ञापकत्वात्, ज्ञापकस्य धूमादेरेतद्रूपं यत्सम्बन्धग्रहणापेक्षं स्वज्ञाप्यज्ञापकत्वम्, तद्योग्यतादयस्तु प्रत्यक्षसामग्र्यन्तर्गतत्वान्न व्युत्पत्त्यपेक्षा भवन्ति शक्तिस्तु नैसर्गिकी यथा रूपप्रकाशिनी दीपादेस्तथा शब्दस्यार्थप्रतिपादने, तस्मान् न समयमात्रादर्थप्रतिपत्तिः ।

अपि चाभिधानभिधेयनियमनियोगरूपः समयो ज्ञानमेव न ततोऽर्थान्तरं ज्ञानं चात्मनिवर्त्तते न च शब्दार्थयोरिति न तयोः सम्बन्धः स्यात्, किं च समयः क्रियमाणः प्रत्युच्चारणं वा क्रियते प्रतिपुरुषं सर्गादौ वा सकृदीश्वरेणेति, प्रत्युच्चारणं प्राक्तन एव क्रियते नूतनो वा, नवस्य तावत्क्रियमाणस्य कथमर्थप्रत्यायनसामर्थ्यमवगम्यते तदवगतौ वा किं तत्करणेन, पूर्वकृतस्य तदा कृतत्वादेव पुनः करणमनुपपन्नम्, एकस्य वस्तुनो ज्ञाप्तिरसकृदावर्त्तते नोत्पत्तिः, प्रतिपुरुषमपि सम्बन्धो भिन्नोऽभिन्नो वा क्रियते, भेदपक्षे कथमेकार्थसंज्ञानं गोशब्दस्य सास्नादिमानर्थः केसरादिमानश्वशब्दस्येति, अभेदेऽपि तथैव कृतस्य करणायोगाज्ज्ञानमेव सम्बन्धस्य न करणम्, सर्गादावपि सकृत्सम्बन्धकरणमयुक्तं तथाविधकालासम्भवादेव न हि शब्दार्थव्यवहाररहितः कश्चित्काल उपपद्यते, तस्मान्नित्यस्यैव सम्बन्धस्य लोकतो व्युत्पत्तिः न पुनः करणम्, व्युत्पत्तिपक्षं च न करणपक्षाभिहिता दोषाः स्पृशन्ति प्रत्यक्षसिद्धत्वात्, प्रत्यक्षं हीदमुपलभ्यते वृद्धानां हि स्वार्थे व्यवहरमाणानामुपश्रृण्वन्तो बालास्ततस्ततः शब्दात्तं तमर्थं प्रतियन्ति, तेऽपि वृद्धा यदा बाला आसंस्तदाऽन्येभ्यो वृद्धेश्यस्तथैव प्रतिपन्नवन्तस्तेऽप्यन्येभ्य इति नास्त्यादिसंसारस्येति ।

अपि च समयमात्रशरणः शक्तिशून्यः शब्दः कथमक्षिनिकोचहस्तसंज्ञादिभ्यो भिद्येत, स हि तदानीं कशाङ्कुशप्रतोदाभिघातस्थानीय एव भवेत्तथा च शब्दार्थं प्रतिपद्यामहे इति लौकिको व्यपदेशो बाध्येत, समयादर्थं प्रतिपद्यामहे इति स्यात्, समयपक्षे च यहच्छाशब्दतुल्यत्वं सर्वशब्दानां प्राप्नोति येन गवाश्वादीनां शब्दानां नियतविषयत्वं न स्यात् ।

यत्पुनरुच्यते जातिविशेषे चानियमात्समयरूपः सम्बन्ध इति, जातिशब्दे नात्र देशो विवक्षितः, किलक्व चिददेशविशेषे कश्चिच्छब्दो देशान्तरप्राप्तप्रसिद्धमर्थमुत्सृज्य ततोऽर्थान्तरे वर्त्तते, यथा चौरशब्दस्तस्करवचन

ओदने दाक्षिणात्यैः प्रयुज्यते, एतच्च समयपक्षे युज्यते, नित्ये तु सम्बन्धे कथं तदर्थव्यभिचार इति, तदप्ययुक्तम्, सर्वशब्दानां सर्वार्थप्रत्यायनशक्तियुक्तत्वात् क्वचिद्देशे केनचिदर्थेन व्यवहारः, अत एव चानवगतसम्बन्धे श्रुते सति सन्देहो भवति कमर्थं प्रत्याययितुमनेनायं शब्दः प्रयुक्तः स्यादिति, असत्यां हि शक्तौ अकृतसमये निरवलम्बना प्रत्यायकत्वाशंकेति, अथ वा ऽऽर्यदेशप्रसिद्ध एव शब्दानामर्थः इतरस्तु म्लेच्छजनसम्मतो नादरणीय एव, तस्मात्समयपक्षस्यातिदौर्बल्यादकृत्रिम एव शब्दार्थयोः सम्बन्ध इति न तत्र पुरुषस्य प्रभविष्णुता ।

* RELATION BETWEEN A WORD AND ITS MEANING

(This section is a dialogue between Mīmāṁsakas who put the question and Naiyāyikas who give the answer)

It is said (by the Mīmāṁsakas) that the relation between a word and its meaning does not depend on the persons. The word possesses an innate expressive power just as the power of burning is inherent in fire. The comprehnsive learning of words with reference to its meaning is derived from the use of experienced linguistic adults. What would a person do ? Such a view is not logically sound. Because unless and untill a convention 'that this word referentially relates to this object' is set up by a person the comprehension of the meaning from a word is not possible.

The Mīmāṁsakas contend that since there is no relation betwen word and its meaning why is it considered at all whether or not the said relation is dependent or independent of a person ? The relation between a word and its meaning is not like the relation 'conjunction' which is realised between a plum and a small reservoir of water. This is not perceived like the relation 'inherence' which subsists between threads and a peice of cloth. No other relations which are based upon any of these two are possible. It has been said that a word is uttered in the mouth but an abject is seen on the ground. The relation of a word with its meaning is also not inferred. When the words like 'razor' and 'sweet ball' etc., are uttered niether a cut in mouth by a razor nor the filling up of mouth by a sweet ball is experienced. Neither the object referred to by a word

*** This chapter is taken from Nyāya-mañjarī, (Āhnika, 4)**

belongs to the place where the word is uttered nor a word belongs to the place where the object exists. The word and the object do not co-exist because the place of articulation, instrumental cause and efforts the condition of sound are not perceived where the object such as 'pot' etc., exist. Moreover, pervasiveness of sound (i.e. a word is more pervasive than sound) is denied.

The Naiyāyikas say that we do not admit a relation which points to contact between word and object. (The Mīmāṁsakas ask) Does it point to the relations like, 'cause-effect, condition-conditioned' or 'resort - resorted' ? (The Naiyāyikas answer) None of these relations hold between them. (The Mīmāṁsakas remark) This shows that there is no relation whatsoever between a word and the object. (The Naiyāyikas reply) it is not the case that there is no relation between a word and the object since a word invariably expresses an object as smoke points to fire. (The Mīmāṁsakas question) Is it the relation 'invariable concomitance' that hold between a word and the object ? (The Naiyāyikas reply) No. Because it would lead to inferential knowledge of word if we subscribe to this view. (The Mīmāṁsakas ask) What is this relation ? (The Naiyāyikas reply) This relation is a conventional one. What is this relation ? It is a relation determined by an act of (expressive) will that hold between the expression and expressibility. (The Mīmāṁsakas raise another question) If you subscribe to the hypothesis of convention why do you refer to conjunction and refute it. (The Naiyāyikas say) This relation holds good from the standpoint of Grammarians because they admit the relation of identity in difference between a word and its meaning. Such a view is not tenable sicnce the convention is set up by a human being and no law is imposed upon the external objects by the mere will of a person. Moreover, the law of nature cannot be changed by the will of a person. Can the object not be expressive of itself ? No. It is not a case that the person having no desire for knowing fire does not know it from smoke. Although he has desire for knowing water yet he cannot but infer fire in spite of his adverse desire.

The relation between smoke and fire is invariable concomitance which is natural one. But in order to determine this relation repeated observations of the co-existence of smoke and fire are necesssary. Similarly, the relation between word and meaning is natural one. But for ascertaining the expressive power of a word we are to depend upon the usages of senior persons and some other factors like traditional meaning of the adjacent word. If it is a natural relation does it behave like that which is realised between light, the illuminator (the lamp) and the object illuminated by it ? Does it make one conscious of this relation when he sees an object illuminated by it ? If it is the case why should one resort to the usages of seniors? A word is merely an indicator. The nature of an indicator like smoke is that it reveals the object to be indicated by it to a person only when he knows the relation between them. The compatibility and such other factors of this relation are included in the condition of perception. But these are not necessary factors for ascertaining usages. The expressive power is as natural as the power of revealing belonging to a lamp. A word is expressive of a meaning by means of this power. It does not resort merely to the conventional relation.

(The Mīmāṁsakas contend) A convention means nothing but a relation realised through expression and expressibility and it is a kind of knowledge belonging to soul. It neither belongs to word nor to an object. It cannot, therefore be relation between word and object. The Naiyāyikas hold that the convention has been set up by God. (The Mīāṁsakas agrue :) Does God repeatedly make it at the dawn of creation for each person wherever each person utters it ? Or, does He frame it once for all persons in the beginning of this world? Does the convention which is repeatedly created for each person becomes new or remain old one at all instances of utterance ?

If it is a new convention then how it helps to communicate the meaning ? If it is already known what is the use of setting a new convention ? That which has been done before is being done again. Since it is done there is no need of repeating it again. An object is known again and again but it is not created more than once. A word may be uttered by the different persons. Does the relation between word and its meaning become different or identical ? If the relation is distinct then how does a word express the same meaning and the word cow signify an animal having dwelap ? How does the word horse refer to an animal having manes ? If the relation is one and the same then the relation which has been created can not be created again. The knowledge cannot be the cause of such relation. It is also not reasonable to hold that the relation has been created by God in the beginning of creation. It is highly impractical to imagine such a time. It does not hold good to conceive a time when there was no verbal transaction.. It is therefore concluded that the relation is eternal and determined from the usage of people. Such a relation is not created. The deffect which undermines this stand that the relation is created does not affect the hypothesis that the eternal relation is learnt since the second one is based upon perceivable reasoning. It is also perceived that the boys learn from the use of seniors that this word refers to this object and so on. When the old persons were boys they learnt it from seniors and so on *ad infinitum*. Thus there is no beginning of this world.

Moreover, how the word which resorts to convention and which lacks intrinsic power of expression is distinguished from winking and signalling by hand? Such being the case a word will be as good as the lashings of a whip or piercing with a hook or beating with a sharp instrument. Then the contention that we understand word-meaning would be impeded. It would be like this that a meaning follows from a convention.

If it is accepted then all words would be reduced to the state of arbitrary ones and words like cow and horse etc. would have no definite meaning.

The Mīāṁsakas anticipate the objection from the Naiyāyikas that the relation between word and meaning is conventional type, because a word does not uniformly refer to the same meaning in all countries. The term *Jāti* conveys a country. But the word *jāti* is not used here in the sense of country. Because a word refers to a definite meaning in particular country. In some other countries the very word does not refer to the same meaning. It points to something else. For instance, the word *caura* which refers to thief is used to mean rice by the people of the south. Such being the case it amounts to saying that the relation between word and meaning is established by the convention. If the relation were eternal such a charge in the meaning of a word would not bave been possibe.

This objection does not stand to reason. Each word has the potency of expressing all possible meanings. A particular word is employed to express a particular meaning in a particular country. When the relation is not known one cannot understand the meaning of a word because a doubt arises as to the determination of exact meaning. If it is admitted that a word does not possess expressive power then no doubt as to its meaning would arise on hearing a word since a conventional relation has not been established or , the meaning of a word which has been accepted by the Aryans should be admitted and that which has been accepted by Non-Aryans should be disregarded. Since the hypothesis that the relation is established by the convention resorts to weak reasons it is concluded that the relation between a word and its meaning is not established by human agency, i.e. it is not artificial. The relation is, therefore, not fashioned by human agency.

* शब्दार्थसम्बन्धनित्यतानिराकरणम्

अत्रोच्यते - न नित्यः समबन्ध उपद्यते, शब्दवदर्थवच्च तृतीयस्य तस्य प्रत्यक्षादिना प्रमाणोनाप्रतीयमानत्वात्।

ननु शक्तिरूपः सम्बन्ध इत्युक्तं शक्तिश्च तदाश्रितेति कथं धर्म्यन्तरवत्पृथक्तया प्रतीयते, नैतत्साम्प्रतम्, स्वरूपसहकारिव्यतिरिक्तायाः शक्तेः सूक्ष्मायाः प्रागेव विस्तरतः प्रतिक्षिप्तत्वात्, न च शक्तिः प्रत्यक्षगम्या द्रव्यस्वरूपवदनुपलम्भात्, नानुमेया कार्याणामन्यथाऽपि घटमानत्वात्, कल्पयित्वा च शक्तिमपरिहार्यः समयः समयमन्तेरणार्थप्रतिपत्तेरसिद्धेः,सिद्धे च समये तत एवार्थसिद्धेः किं नित्यसम्बन्धाश्रयणेन।

यत्तूक्त समयस्यं परुषेच्छाधीनत्वात्तस्याश्चाव्याहतप्रसरत्वाद्वाच्यवाचकव्यत्ययः स्यादिति तदयुक्तम्, शक्त्यभावे शब्दस्यैव वाचकत्वे योग्यत्वात्, का पुनः शक्त्यभावे योग्यताऽस्येति चेद् योऽयं गत्वादिजातियोगः क्रमविशेषोपकृतः गत्वौत्वादिसामान्यसम्बन्धो हि यस्य भवति स वाचकत्वे योग्य इति इतरस्तु वाच्यत्वे, यथा द्रव्यत्वाद्यविशेषेऽपि वीरणत्वादिसामान्यवतां च पटनिष्पत्तौ न च तत्र शक्तिरस्तीत्युक्तम्, न च कारणे कार्यं सदिति सांख्यैरिव भवद्भिरिष्यते तस्यामसत्यामपि शक्तौ सामान्यविशेषसम्बन्धस्य नियामकत्वान्न वाच्यवाचकयोर्व्यत्यय इति न शक्तिरूपः शब्दोर्थयोः सम्वन्धः, न च तयोरविनाभावे धूमाग्न्योरिव सम्बन्धः, तत्र हि सम्बन्धः प्रतीयमान एव प्रतीयते धूमोऽग्निं विना न भवतीति इह पुनरयमस्मात्प्रतीयते इति एतावदेव व्युत्पत्तिपर्यवसानम्, अत एवावगतिपूर्विकैवावगतिरिहेत्यनुमानाच्छब्दस्य भेद उक्तः, प्रकाशकत्वमपि शब्दस्य ससमयप्रसादोपनतमेव न स्वाभाविकम्, सांसिद्धिके हि तथात्वे भ्रमित्वादिप्रयुक्तादन्यतो वा यतः कुतश्चिदभिनवादपि दीपादिव शब्दार्थप्रतीतिः स्यात्।

यत्तु नैसर्गिकेऽपि प्रकाशकत्वे शब्दस्य धूमादेरिव ज्ञापकत्वात्संबन्धग्रहणसापेक्षत्वमुक्तं स एष विषम उपन्यासः, न हि धूमादेः प्रत्यायकत्वं स्वाभाविकम् अनलाविनाभावित्वं तु तस्य निजं बलं तत्र चागृहीते तस्मिन्प्रतीतिरेव न जायते इति युक्तं तद्ग्रहणं प्रतीत्यर्थम्, इह तु प्रतीतिशक्तिरेव

---

* न्यायमञ्जरीत उद्धृतसिदं प्रकरणम् (न्या. म. ४.)

स्वाभाविकी भवता ऽभ्युपगम्यते सा चेत्स्वाभाविकी किं व्युत्पत्यपेक्षणेनेति, यच्चोच्यते प्रत्यायक इति प्रत्ययं दृष्ट्वाऽवगच्छामो न प्रथमश्रवण इति यावत्कृत्वःश्रुतेनेयं संज्ञाऽयं संज्ञीत्यवगम्यते तावत्कृत्वः श्रुतादर्थावगम इति सोऽयं समयोपयोग एव कथितो भवति, संज्ञासंज्ञिसम्बन्धो हि समय एवोच्यते तदुपयोगमन्तरेण प्रत्यायकत्वानवगमान् न स्वाभाविकी शक्तिः।

यत्त्वभ्यधायि समयस्य ज्ञानात्मकत्वादात्मनि वृत्तिः न शब्दार्थयोरित्येतदप्यचतुरश्रम्, तदाश्रयत्वाभावेऽपि ज्ञानस्य तद्विषयत्वोपपत्तेः।

यदप्यभाणि समयमात्रशरणे सृणिप्रतोदनोदननिर्विशेषे शब्दे शब्दादार्थं प्रतिपद्यामहे इति व्यपदेशो न स्यादिति तदपि न किंचित्, नैसर्गिकशक्तिपक्षे अपि शक्तेरर्थं प्रतिपद्यामहे न धूमादिति स्यात्, तदङ्गत्वादविनाभावादेर्न तथा व्यपदेश इति चेत् तदितरत्रापि समानम्

धूमे हि व्याप्तिपूर्वत्वं शब्दे समयपूर्वता।
नानयोस्तदपेक्षायां करणत्वं विहन्यते।।

अपि च, लौकिको व्यपदेशः समयपक्षसाक्षितामिव भजते देवदत्तेनोक्तम् अमुतः शब्दादमुमर्थं प्रतिपद्यस्वेति एवं हि व्यपदिशति लोकः, तस्मात्समय एव, अतश्चैवं देशान्तरे सङ्केतवशेन तत एव शब्दादर्थान्तरप्रतिपत्तिः।

नन्वत्रोक्तं सर्वशब्दाः सर्वार्थप्रत्यायनयुक्ता इति केनचिदर्थेन क्वचिद्व्यवहार इति तदेतदयुक्तम्, शक्तीनां भेदाभेदविकल्पानुपपत्तेः, न शब्दस्वरूपाद्भिन्नाः शक्तयः तथाऽनवभासात् अव्यतिरेके चैकस्माच्छब्दादनन्यत्वात्परस्परमव्यतिरेकस्तासां स्यात्, न च भिन्नकार्यानुमेया भिन्नाः शक्तयः कार्यभेदस्यान्यथा ऽप्युपपत्तेः, सर्वशक्तियोगे च सर्वार्थप्रत्ययप्रसगङ्, समयोपयोगो नियामक इति चेत् स एवास्तु किं शक्तिभिः।

यदप्यगादि शब्दश्रवणे सति सर्वार्थविषयसन्देहदर्शनात्सर्वत्र तस्य शक्तिः कल्प्यते इति तदप्यसारम्, न हि शक्तिकृतः सन्देहः किं तु गत्वादिवर्णसामान्यनिबन्धनः तथा च गत्वादिजातिमतां वर्णानामर्थे वाचकत्वमवगतम्, अमी तज्जातियोगिनो वर्णाः कस्यार्थस्य वाचकाः स्युरिति भवति सन्देहः।

यत्पुनरवादि स एव शब्दस्यार्थो यत्रैनमार्याः प्रयुञ्जते न म्लेच्छजनप्रसिद्ध इति तदेतत्कथमिव शपथमन्तरेण प्रतिपद्येमहि, न हि म्लेच्छदेशेऽपि तदर्थप्रत्ययो न जायते,

बाध्यते वा सन्दिग्धो वेति कथं न बाधिका, अक्षादिवच्च विकल्पमानार्थोपपत्तेः व्यवस्थितविषय एव विकल्पो भविष्यति, पिकनेमतामरसादिशब्दानां च भवद्भिः म्लेच्छप्रयोगादर्थनिश्चय आश्रित एव, अवेष्ट्यधिकरणे च राज्यशब्दमान्ध्रप्रसिद्धेऽर्थे वर्णितवन्तो भवन्त इत्यलमवान्तरचिन्तनेन, तस्मात्समय एव सम्बन्ध इति युक्तम्, तदुक्तं जातिविशेषे चानियमादिति।

अथ यदुक्तं समयः प्रत्युच्चारणं प्रतिपुरुषं च तत्करणम् अनभ्युपगतमेव दूषितम्, सर्गादौ सकृदेव समयकरणमिति नः पक्षः, अत एव न सर्वशब्दानां यदृच्छातुल्यत्वम्, केषांचिदेव शब्दानामस्मदादिभिरद्यत्वे सङ्केतकरणात्त एव यदृच्छाशब्दा उत्पद्यन्ते, सर्गादिश्च समर्थित एव ईश्वरसिद्धावप्यविकल्पमनुमानमुपन्यस्तम्, एष एव चावयोर्विशेषो यदेष शब्दार्थसम्बन्धव्यवहारस्तवानादिर्मम तु जगत्सर्गात्प्रभृति प्रवृत्त इति, अद्यत्वे तु शब्दार्थसम्बन्धव्युत्पत्तौ तुल्य एवावयोः पन्थाः, तत्रापि त्वयं विशेषो यत्तव शक्तिपर्यन्ता व्युत्पत्तिर्मम तु तद्वर्जमिति, तथा चेयमियती व्युत्पत्तिः लोके दृश्यते यदयमस्य वाच्योऽयमस्य वाचक इति न पुनः शक्तिपर्यन्ता व्युत्पत्तिरस्ति, तथा हि यत्र श्रृङ्.गग्राहिकया शब्दमर्थं च निर्दिश्य सम्बन्धः क्रियते तत्रेयन्तमेवैनं क्रियमाणं पश्यामः, अयमस्य वाच्योऽयमस्य वाचक इति, यत्रापि च वृद्धेभ्यो व्यवहरमाणेभ्यो व्युत्पद्यते तत्रापीयदेवासौ जानाति अयमर्थः अमुतः शब्दादनेन प्रतिपन्न इति न त्वन्यस्य काचिच्छक्तिरस्तीति, इयत्यैव च व्युत्पत्त्या शब्दादर्थप्रत्ययोपपत्तेरस्याश्चापरिहार्यत्वादधिककल्पनाबीजाभावाच्च न नित्यः शब्दार्थसम्बन्धः, अत एव च सम्बन्धस्त्रिप्रमाणक इति "यत्त्वयोच्यते तदस्माभिर्न मृष्यते, "शब्दवद्धाभिधेयांश्च प्रत्यक्षेणात्र, पश्यति"' इति सत्यं "श्रोतुश्च प्रतिपन्नत्वमनुमानेन चेष्टया" इत्येतदपि सत्यम्, "अन्यथाऽनुपपत्त्या तु बुध्येच्छक्तिं द्वयात्मिकाम्" इत्येतत् न सत्यम्, अन्यथाऽप्युपपत्तेरित्युक्तत्वात्, तस्माद् द्विप्रमाणकः सम्बन्धनिश्चयो न त्रिप्रमाणकः, तदेवं शब्दस्य नैसर्गिकशक्त्यात्मकसम्बन्धाभावादीश्वरविरचितसमयनिबन्धनः शब्दार्थव्यवहारो नानादिः।

नन्वीश्वरोऽपि सम्बन्धं कुर्वन्नवश्यं केनचिच्छब्देन करोति तस्य केन कृतः सम्बन्धः, शब्दान्तरेण चेत्तस्यापि केन कृतस्तस्यापि केनेति न कश्चिदवधिस्तस्मादवश्यमनेन सम्बन्धं कुर्वता वृद्धव्यवहारसिद्धाः केचिदकृतसम्बन्धा

एव शब्दा अभ्युपगन्तव्या अस्ति चेद् व्यवहारसिद्धिः किमीश्वरेण किं वा तत्कृतेन समयेनेत्यनाद्यपक्ष एव श्रेयान्, उच्यते–

अस्त्रमायुष्मता ज्ञातं विषयस्तु न लक्षितः.।
अस्मदादिषु दोषोऽयमीश्वरे तु न युज्यते ।।
नानाकर्मफलस्थानमिच्छयैवेदृशं जगत् ।
स्रष्टुं प्रभवतस्तस्य कौशलं को विकल्पयेत् ।

इच्छामात्रेण पृथिव्यादेरियतःकार्यस्य करणमस्मदादीनां यन्मनोरथपदवीमपि नाधिरोहति तदपि यतः सम्पद्यते तस्य कियानयं प्रयासः, तदत्रेश्वरसद्भावे परं विप्रतिपत्तयः, तस्मिस्तु सिद्धे क्व एवं विकल्पानामवसरः, उक्तं च तत्सिद्धौ निरपवादमनुमानम्, वयं तु न कर्त्तार एव सम्बन्धस्य यत एवमनुयुज्येमहि--

अङ्.गुल्यग्रेण निर्दिश्य कंचिदर्थं परः स्थितम् ।
व्युत्पादयन्तो दृश्यन्ते बालानस्मद्विधा अपि ।।

तस्मादीश्वरविरचितसम्बन्धाधिगमोपायभूतवृद्धव्यवहारलब्धतद्व्युत्पत्तिसापेक्षः शब्दोऽर्थमवगमयतीति सिद्धम्, न च नित्यसम्बन्धाभावेऽपि शब्दस्यार्थास्पर्शित्वं समयबलेनार्थप्रत्ययस्याबाधितस्य सिद्धेरित्युक्तत्वाद् इत्यलं विस्तरेण।

तस्मात्पदे च वाक्ये च सम्बन्धे च स्वतन्त्रता ।
पुरुषस्योपपन्नेति वेदानां तत्प्रणीतता ।।

*REFUTATION OF ETERNITY OF THE RELATION BETWEEN WORD AND MEANING

The fact that the relation between word and meaning is eternal does not seem appropriate. Because the relation as a distinct entity is not cognised in the same manner as the word and meaning are, conceived by the valid means of knowledge.

The Mīmāṁsakas contend that the expressive power is stated as the relation between word and meaning. The very power is dependent one. How can it be cognised like an independent whole ? This contention does not hold good. Because the existence of a subtle power which is exclusive of a cause and its auxiliary conditions has been refuted earlier. The expressive power is not comprehended through perception. Since it does not possess a form like substance, it also cannot be inferred. Because the effect may be brought about by a cause other than the expressive power. Although one assumes a relation yet it remains unavoidable. Had the relation not existed the meaning of a word would not have been ascertained. If the relation is established then the meaning of a word is determined. As such what is the necessity of assuming an eternal relation ?

The Mīmāṁsakas say that as a convention depends on the desire of human being and the force of desire is irresistable so there is possibility of the converse of expressibility. Such an apprehension does not stand to reason. Because a word is competont to be expressive of a meaning without a power. The opponent objects that what is this competence without a power ? The reply is

*** This chapter has been taken from Nyāya-mañjarī, (āhnika 4.).**

that the word *go* which contains the universal like *gatva* and *autva* etc. in proper succession is entitled to be expressive. But the object which is referred to by it is expressible. For instance, *vīraṇa* 'a kind of grass' is a substance. There are substances which do not belong to the class of *vīraṇa*. Of all substances *vīraṇas* alone produce a mat because the universal *vīraṇatva* inheres in them. You do not subscribe to the views of the Saṃkhyist who hold that an effect in its potential from belongs to its cause. The *śakti* as a power of producing effect cannot be taken as the relation between word and meaning. Because there is no converse of relation between expressive unit and the expressible even if there is no such *śakti* owing to the fact that there is a fixed norm as regards the particular and general relation. The relation is not that of invariable concomitance holding between smoke and fire. This relation appears in a form of indication, i.e. when this relation is cognized we know that smoke does not exist at a spot where fire is absent. Here also we know that the object is known from a word. This is the final lesson that we learn from experience. It has been stated that verbal knowledge is different from the inferential one because the former one presupposes the knowledge of expressive relation. The fact that word reveals the object does not point natural process but to something which is undermined through the impact of convention. If it is held that a word reveals an object by its innate power in the situation when the word is not uttered either by mistake or negligence it would reveal the intended object just as a lamp illuminates the same.

The Mīmāṁsakas hold that the power of revealing is innate in a word and the word reveals its meaning only when the relation between them like that between smoke and fire is known. But those two cases are not parallel.

The smoke has no inherent power of revealing fire. But it is invariable concomitant of fire by nature. If such relation is not known the knowledge of fire does not follow from perception of smoke. You (the Mīmāṁsakas) admit that a word has an innate power of revealing its meaning. If it is innate then why should verbal knowledge depend upon the determination of relation from experience. You say that a word is a revealer only when we comprehend that a word produces cognition and we fail to understand this nature of a word only on hearing it. But as many times as we hear a word and know that this is a name and this is the object named so, so many times we comprehend the meaning of a word heard by us. Thus the potentiality of expressive relation is stated by this. The relation between a name and its referent is called *samaya.* Since without the efficacy of the expressive relation a word cannot convey its meaning, it is resolved that the power of expressing a meaning is not innate.

You have stated earlier that a conventional relation partakes the nature of consciousness and belongs to the soul. But it neither belongs to a word nor to an object referred to by it. This contention is not sound. Because the consciousness does not find a locus in them, but it is related to word and meaning which form its content.

You have also said that if a word expresses it meaning only by means of its conventional relation then the word will be better in no way than the piercing of a hook or the lashing of a whip. Such a criticism is not tenable, if the power of conveying is inherent in the word it should have been held that the power itself conveys a meaning but not the word. If the Mīmāṁsakas say that the power of conveying like concomitance is mere a condition of conveying the meaning then same logic would be applicable to other cases.

As smoke depends on the relation of invariable concomitance so the word depends on the relation of convention. The causehood of both, however, is not affected in spite of such dependence.

Moreover, coloquial expression bears the evidence of conventional relation. For instance, Devadatta has said, Learn this meaning from this word' An ordinary person talks like this. The conventional relation should, therefore, be accepted. It is the conventional relation which alone explains why a particular word conveys different meaning in another country.

You (the Mīmāṁsakas) have said that each word is capable of expressing all sorts of meaning. Some words are restricted to particular meanings in different countries. This contention is not sound. Because the expressive powers cannot be said as identical or distinct. It cannot be said also that since the characters of words are different the expressive powers are different. This is not presented to our consciousness. If it is admitted that the said power is identical with a word then all the different powers belonging to a word should also be identical. If it is held that different powers are inferred from different effects then it is also not sound. How can one account for the diversity of effects as due to difference in powers? The variety of effects may also be otherwise explained. If it is contended that all words are capable of conveying all meanings then a word should express different meanings. If you say that it is restricted to particualr meaning by means of conventional relation and if this is the contention (that the conventional relation is an important factor) then there is no need of assuming so many powers.

You contend that when word is heard it leads to a doubtful state since many meanings flash in the mind. It prompts one to assume that a word possesses multifarious powers. But this is not sound. This doubt does not resort to power. This pertains to universals belonging to letters like *gatva*. A word consisting of many letters each of which has particular universal is expressive of meaning. If none of its constituents is expressive of a meaning then how can a word which is an aggregate of letters having no meaning express a meaning ? This raised the doubt.

It has been also said that the meaning of a word is that which is conveyed by it used by the Aryans but not that which has been accepted by non-Aryans. How can one follow this if they donot take an oath to prove the fact. In other words, it does not hold good. It is not a fact that a word does not express its meaning in non-Aryan country. It is also not true that the said meaning is not contradicted or porduces doubt. Why would then this be no real ? It cannot be said that meaning of an Aryan word contradicts the non-Aryan meaning. Because the word like *akṣa* admits alternative meaning. The scope of each alternative meanings will be well defined and fixed up. You have learnt the meanings of the word *pika* 'cuckoo' *nema* 'portion' and *tāmarasa* 'lotus' etc. from non-Aryan sources. In the section of *āveṣṭi* i.e. *rājasūya* sacrifice you have explained the word *rājya* 'state' following the description of non-Aryan. There is no need of discussing minor details. It is thus deducted that the relation between word and meaning is conventional one. It should also be borne in mind that a word referring to universal has no fixed meaning.

The Mīmāṁsakas have said that whenever a person utters a word he establishes each time a new convention. But they have not comprehended our point. What has been criticized by them is not our contention. Our contention is that God has created a convention once for all at the outset of creation. It is not the fact that the meaning of all words has been determined arbitrarily. The meanings of some words have been fixed up by us at our discretion. These are arbitrarily fashioned. We have established the beginning of creation along with the existence of God by an irrefutable syllogistic evidence. The point of difference between us is that you (the Mīmāṁsakas) hold that the expressive relation is eternal but we contend that the relation as fixed at the begining of this creation is same till now. But at present, our way of ascertaining the relation between word and meaning appear alike. Even then it is the difference that you say that the expressive relation is ascertained as inherent in word. But we hold that expressive relation is a exclusive notion. It is found in the linguistic community that this is the conveyed meaning of this word and this word is expressive of this meaning. But no use is found as regards the meaning being an innate power of word. It has been stated that the relation is specifically mentioned. This much may be admitted that this is expressed as the meaning of this word and this word is expressive of this meaning. When a person learns the use of words from the usage of persons adept in the use of words he knows only this that this word is expressive of this meaning. But he does not know about the innate power of a word. With this it is establisehd that the word is expressive of a meaning. The relation between word and meaning is not eternal. The contention of the Mīmāṁsakas that the expressive relation is established by three kinds of proof is not accepted by us. It is true that a word is learnt form the behaviour of an experienced person who command and of another such persons who carry out the same from the knowledge of the command by way perception. It is also true that listener rightly infers from the movement of the experienced person who has been at the disposal of the employer to obey his command. But it is not true that the power of expression which rests upon a word and its meaning is a presumption. It has already been stated that this would be effected by other ways. It thus may be concluded that the relation is determined by two kinds of proof but not three. It should be noted that a word does not posess innate power, and lacks eternal relation to its meaning. But God has established a relation between word and its meaning at his discretion. The linguistic behaviour owes its origin to God and hence is not beginingless.

The Mīmāṁsakas say that when God sets up relation he does it by means of words. How are these known to be related to meaning ? If these are so related then how is their meaning communicated to others by means of different set of words ? If it is held that God has connected words with their meanings then there will be a *regressus ad infinitum* to convey the meaning of a word by means of never ending series of words of unknown meanings. If it is admitted that God has created relation which has been attested by the usage of experienced predeccessors and if a word is known to be related to its meaning by means of usages of experienced person then what is the need of admitting the relation as created by God ? It is better to admit the first stand that the expressive relation is eternal.

The Naiyayikas answer as follows :

Your goodself has learnt something but not marked the subject to be aimed at. The objections are applicable only to us. But they donot apply to God.

God, who has created the universe by his mere will for the fruition of the diverse actions has a skill which remains beyond the ken of our imagination i.e. who is able to alter his skill.

By mere act of will we cannot conceive the frame-work of this world and such other objects. It does not amount to difficulty on the part of God. There may be counter agruments which disprove the existence of God. If the existence of God is established once then there is no scope for alternatives. We have already shown our irrefutable arguments which prove the existence of God. Since we donot create the relation we should not be taken to task.

Men like us intiate children into words by pointing finger-ends to the yonder objects.

It is therefore to be admitted that the linguistic transactions of the old persons constitute the means of expressive relation introduced by God. Thus processed through experience of seniors the word communicate the ideas. Although the relation between word and meaning is not eternal yet the word does not remain unrelated to the meaning. On the strength of expressive relation set up by God the uncontradicted knowledge of the meaning follows from a word. Since we have already discussed it there is no need of dilation.

It is therefore, sound to hold that the independence which amounts to subsist in word, in sentence and in relation resorts to God, i.e. God has introduced and established the relation between word and meaning. This proves the hypothesis that God is the author of the Vedas is reasonable.

*गवादिशब्दार्थविचारः

(जात्याकृत्योर्भेदाभेदचिन्ता)

तत्राकृतिपदेनेह संस्थानमभिधीयते ।
सूत्रे पृथगुपादानान्न जातिर्जैमिनीयवत् ।।

जैमिनिर्हि आकृतिस्तु क्रियार्थत्वादित्याकृतिं जातिमुपदिशति, भाष्यकारोऽपि सास्नादिविशिष्टाकृतिरिति ब्रुवाणः तथैव व्यवहरति, वार्त्तिककृता हि तद्व्याख्यातम्–

जातिमेवाकृतिं प्राहुर्व्यक्तिराक्रियते यया ।
सामान्यं तच्च पिण्डानामेकबुद्धिनिबन्धनमिति ।।

इह तु व्यक्त्याकृतिजातयस्तु पदार्थ इति सूत्रकारो जातेः पृथगाकृतिवचो वदति संस्थानमेवाकृतिं मन्यते, लोकोऽपि अवयवसन्निवेशात्मिकामाकृतिं व्यपदिशति "यत्राकृतिस्तत्र गुणा वसन्ती" ति, तस्मादवयवसन्निवेश एवाकृतिरुच्यते ।

तस्याश्च शब्दार्थतोपपद्यते न वेति परीक्षार्हत्वमस्त्येव न जैमिनीयवदुपेक्षितुं सा युक्तेति, तद्व्यक्त्याकृतिजातिसन्निधौ प्रयोगाद् गोशब्दस्य कोऽर्थ इति विचार्यते ।

व्यक्तिमात्रस्य पदवाच्यत्वनिरूपणम्

तत्राकृतिवादिनस्तावदाहुः प्रयोगप्रतिपत्तिभ्यां किल शब्दार्थनिश्चयः, वृद्धाः स्वार्थेन व्यवहरन्तो यस्मिन्नर्थे गोशब्दं प्रयुञ्जते श्रोतारश्च यमर्थं ततः प्रतिपद्यन्ते स तस्यार्थः, तत्र यदि गोशब्दः केसरादिमति न प्रयुज्यते सास्नादिमति च प्रयुज्यते तदसाधारणसन्निवेशविषय एवावगम्यते, प्रत्यक्षविषये गौरित्यादि पदं प्रयुज्यते प्रत्यक्षं चाकृतिविषयमश्वपिण्डसन्निवेशाद्विलक्षणो हि गोपिण्डसन्निवेश इन्द्रियेण प्रतीयते तत्कृतमेव वस्तुष्वितरेतरवैलक्षण्यम्, अतः प्रत्यक्षविषये पदं प्रवर्त्तमानमाकृतावेव वर्तितुमर्हति, प्रेषणादिक्रियायोगश्च व्यक्तिद्वारक

* न्यायमञ्जरीत उद्धृतमिदं प्रकरणम् (न्या. म. ५.)

आकृतेर्भविष्यतीति तदुक्तम् –

तदयुक्तं प्रतिव्यक्ति भिन्नसंस्थानदर्शनात् ।
आनन्त्यव्यभिचाराभ्यां संबन्धज्ञप्त्यसम्भवात् ।।

न नियतस्य शबलेयसन्निवेशस्य गोशब्दो वाचकः तदभावेऽपि बाहुलेयसन्निवेशदर्शनात्, न च त्रैलोक्यान्तर्गतसकलगोपिण्डसन्निवेशवचनत्वमनुगतं शक्यम् आनन्त्यात्ततश्च नाकृतिः शब्दार्थः तस्यां क्रियाऽनुपपत्तेः न हि प्रेषणा दिक्रियासाधनं सन्निवेशोऽपि तु व्यक्तिः।

न च गामानयेत्युक्तः सत्यामपि तथाकृतौ ।
चित्रमृत्स्नामयीं कश्चिद्गामानयति बुद्धिमान् ।।

ननु जातिवाच्यत्वपक्षेऽपि गोत्वजातेः सर्वगतत्वात् किमिति मृद्गवानयनं नानुष्ठीयते, उच्यते हस्ती किं नानीयते सर्वगतत्वाज्जातेः। अथ सर्वत्रास्तित्वेऽपि व्यंजकव्यक्तिनियमेनापहनूयते हन्त तर्हि सास्नादिमत्प्राणी गोत्वजातेरभिव्यंजको न मृद्गव इति नातिप्रसङ्गः, सन्निवेशस्य च तत्र भावात्तद्वाच्यत्ववादिनः नैनमतिप्रसङ् गमतिक्रामन्ति, किं चाकृतिवचनत्वे गोशब्दस्य शुक्लादिगुणवाचिभिः पदान्तरैः सामानाधिकरण्यं न प्राप्नोति न हि शुक्लादिगुणा आकृतिवृत्तयः तत्सामानाधिकरण्यादिवलवत्तया वरं व्यक्तिः शब्दार्थ इष्यताम्।

व्यक्तौ तावत्क्रियायोगो जातौ सम्बन्धसौष्ठवम् ।
नाकृतौ द्वयमप्येतदिति तद्वाच्यता कुतः ।।
जातिव्यक्त्योरतःकार्या वाच्यत्वे संप्रधारणा ।
तत्र व्यक्त्यभिधेयत्ववादिभिस्तावदुच्यते ।।

प्रयोगचोदनासामञ्जस्याद्व्यक्तिः शब्दार्थः आलम्भनविशसनप्रोक्षणादिचोदना जातावसंगता भवन्ति। न हि जातिरालभ्यते विशस्यते प्रोक्ष्यते वा अपि च षड् देया द्वादश देया चतुर्विंशतिर्देया इति न जातिः षडादिसंख्याभिर्युज्यतेऽपि तु व्यक्तिः तस्मात्सैव शब्दार्थः, अपि च यदि पशुरुपाकृतः पलायेतान्यं तद्वर्णं तद्वयसमालभेतेति यदि जातिः शब्दार्थः स्यादन्यस्यालम्भो नावकल्पेतान्यस्यापनीयमानस्य पशुद्रव्यस्य सैव जातिः तत्रान्यत्वसंबन्धो व्यक्तेरवकल्पते न जातेरित्यतोऽपि व्यक्तिः शब्दार्थः।

चयापचयसंघातस्वस्वामित्वादिकल्पनाः ।
यान्ति व्यक्त्यभिधेयत्वपक्षे झडिति संगतिम् ।।
न व्यक्तिलक्षणद्वारमियत्कार्यं च युज्यते ।
वक्रः पन्था न गन्तव्यः प्रष्ठे वहति वर्त्मनि ।।
उपलक्षणमाश्रित्य जातिसंबन्धवेदनम् ।
प्रसेत्स्यतीति नानन्त्यव्यभिचारकृतो ज्वरः ।।

किंच –

प्रत्यक्षविषये वृत्तिः पदस्येष्टा परैरपि ।
निष्कृष्टं न च सामान्यमात्रं प्रत्यक्षगोचरः ।।
व्यक्तेरेव पदार्थत्वं तस्मादभ्युपगम्यताम् ।
तथा च बुद्धिस्तत्रैव श्रुतशब्दस्य जायते ।

जातिमात्रस्य वाच्यत्वनिरूपणम्

तदेतज्जैमिनीयैर्न क्षम्यते, तथा हि व्यक्तिमात्रं वा शब्दार्थ इष्यते विशिष्टा वा व्यक्तिः, न तावद्व्यक्तिमात्रं न हि यस्यां कस्यांचिद् व्यक्तौ गोशब्दं वक्तारः प्रयुंजते, न च यां कांचिद्व्यक्तिं गोशब्दाच्छ्रोतारः प्रतिपद्यन्ते, अथ गोत्वविशिष्टव्यक्तिः शब्दार्थ इत्युच्यते, गोत्वमेव हि तर्हि गोशब्दार्थो न व्यक्तिः, कथम्, श्रूयतां यदि हि व्यक्तिः शब्दार्थो भवेद् व्यक्त्यन्तरे न प्रयुज्येत अथ व्यक्त्यन्तरेऽपि प्रयुज्यते सर्वव्यक्तिसाधारणस्तस्यार्थो न व्यक्तिः ।

ननु व्यक्त्यन्तरमपि व्यक्तिरेव सोऽयं व्यक्तौ गोशब्दः प्रयुक्तो न सामान्ये, मैवम्, व्यक्तौ चेद् गोशब्दः प्रयुज्यते कर्कादिव्यक्तावपि प्रयुज्येत, यत्र प्रयोगोऽस्य दृष्टस्त्रत प्रयुज्यते इति चेद् अद्य जातायां गवि मा प्रयोजि, न हि तत्र प्रयोगोऽस्य दृष्ट इति तस्माद्दर्शनमकारणम्, प्रतिव्यक्ति तस्यासम्भवात्, व्यक्तौ च शब्दार्थे इयं वा गौरियं वा गौरिति प्रतिपत्तिः स्यान्न त्वियमपि गौरिति भवति चेयं प्रतीतिर्न नायमविद्यमाननियन्तृक एव यदृच्छाशब्दप्रयोगः प्रवर्त्तते इति नियामकमस्य चिन्त्यम्, गोत्वमेव नियामकमिति चेदायुष्मन्साधुबुध्यसे किं तु तद्गोत्वमवगतमनवगतं वेति वक्तुमर्हति ।नानवगतमतिप्रसङ्गादवगतं चेत्कुतस्तदवगच्छामः शब्दादन्यतो, वा नान्यतः प्रमाणान्तरासन्निधानात्, शब्दाच्चेति हि शब्दः प्रथमतरं गोत्वे वर्त्तितुमर्हति नागृहीतविशेषणा विशिष्टे बुद्धिरिति न्यायात् ।

ननु जातिं विशेषणत्वेन व्यक्तिं च विशेष्यत्वेन वक्ष्यति गोशब्दः न शक्नोति वक्तुमतिभारप्रसङ्गात्, न च व्यक्त्यवगतौ गतिरन्या नास्ति यत् इयन्तं शब्दे भारमारोपयेम, न हि वयं व्यक्तिप्रतीतिं भवन्तीमपह्नुमहे नापि भवन्तीं जातिप्रतीतिमपह्नुमहे उभयप्रतीतेः प्रत्यात्मवेदनीयत्वात्, उभयत्र चाभिधात्री शक्तिरतिभारः शब्दस्यान्यतरप्रतीत्या चान्यतरप्रतीतिसिद्धेः, तत्र गोशब्दः किं जातौ वर्त्तमानः व्यक्तिमाहोस्विद्व्यक्तौ वर्त्तमानो जातिमाक्षिपत्विति विचारणायां जाते र्विशेषणत्वात्पूर्वतरं प्रतिपत्तिरिति सैव शब्दार्थो भवितुमर्हति, तस्यां च शब्दादवगतायां तत एव व्यक्त्यवगमः सेत्स्यतीति नोभयत्र शाब्दो व्यापारः।

ननु दण्डिशब्दादिव विशेषणं च जातिं विशेष्यां च व्यक्तिं गोशब्दादेव प्रतिपत्स्यामहे कोऽस्यातिभारः, विषमोऽयं दृष्टान्तः तत्र न हि प्रकृतिप्रत्ययविभागेन द्वयप्रतीतिरवकल्पते। ।दण्डशब्दः प्रकृतविशेषणमभिवदति, मत्वर्थीयप्रत्ययश्च विशेष्यमिति, गोशब्दे तु नैष न्यायः सम्भवति, तत्र न विशेषणे दण्डिशब्दो वर्त्तते न विशेष्ये ।दण्डशब्दः, इह तु गोशब्द एक एव स च विशेषणे विशेष्ये वा वर्त्तेत, विशेष्ये वर्त्तमानो विशेषणे प्रमाणान्तरमपेक्षते विशेषणे तु वर्त्तमानस्तदवगमाय विशेष्यमाक्षिपतीति न कश्चिद्दोषः, तदिदमात्मप्रत्यक्षं यच्छब्दे उच्चरिते व्यक्तिरवगम्यते स किं शब्दादुत जातेरिति विवेको न प्रत्यक्षः स युक्त्याऽवगम्यते, शब्दस्य द्वयाभिधाने यत्नगौरवाद्विरम्य व्यापारस्य चासंवेदनात्, अन्तरेणापि च शब्दं जात्यवगमाद्व्यक्तिप्रतीतिदर्शनात् जातित एवैषा व्यक्तिप्रतीतिः जातिप्रतीतिश्च शब्दादिति निश्चीयते, भवद्भिरपि च विशेषणज्ञानपूर्विका विशेष्यावगतिरङ्गीकृतैव, यथाऽऽह कणव्रतः "समवायिनः श्वैत्याच्छ्व्यैत्यबुद्धेश्च श्वते बुद्धिस्ते कार्यकारणभूते" इति।

यत्पुनरभिहितम् आलम्भनविशसनप्रोक्षणादिक्रियायोगाद्व्यक्तिः शब्दार्थ इति, तदप्यनैकान्तिकम्, जातावपि क्वचित्क्रियायोगदर्शनात् श्येनचितं चिन्वीतेति।

नन्वत्रापि व्यक्तेरेव कार्याङ्गता नामूर्त्ताया जातेः, मैवम्, न ह्यत्र श्येनः साधनत्वेन निर्दिश्यते पशुना यजेतेतिवत्, कर्मण्यग्न्याख्यायामिति त्वभियुक्तस्मरणाच्चयननिर्वर्त्यः श्येन इति शब्दार्थोऽवगम्यते, न च श्येनव्यक्तिश्चयनेन निर्वर्त्तयितुं पार्यते, अग्न्याख्यायामिति स्मर्यते न हीष्टकानिचयेन पतत्री अभिनिवर्त्तते तदाकारस्त्वग्निरभिनिर्वर्त्यते इत्यत्र जातेः क्रियाङ्गत्वं व्यक्तेः।

ननु व्यक्त्या सादृश्यं सम्पादयिष्यते न शक्यते सम्पादयितुं व्यक्तयन्तरवैसादृश्यस्यापि सम्भावाद्यो ह्येकया व्यक्त्या सदृशः सोऽन्यविसदृशो भवति,

यत्त्वमूर्त्तत्वाज्जातेः न क्रियाङ्गत्वमिति नैष दोषः, अमूर्त्तनामपि गुणकर्मणां क्रिया साधनतयोपपत्तेः, अरुणया क्रीणाति अभिक्रामंजुहोतीति, व्यक्त्याक्षेपद्वारेण चालम्भनविशसनप्रोक्षणादिप्रयोगचोदनासु साधनत्वं जातेरुपपत्स्यते।

लक्षितव्यक्तिसाध्यं तु तत्साध्यं कार्यमिष्यते।
यथा भूतेन्द्रियोत्पाद्यमात्मकर्तृकमुच्यते।।

आत्मा तावत्सर्वकर्मस्वधिकृतः कर्त्ता च स चामूर्त्तत्वाद् देहेन्द्रियद्वारेण औदुम्बरीसंमार्जनांज्यावेक्षणदीनि कार्याणि निर्वर्त्तयन् कर्त्ता तेषु भवति, एवं जातिरपि व्यक्तिवर्त्मना तन्निर्वर्त्तयन्ती साधनतां लप्स्यते।

अतश्च जातिरेवाङ्गमिति मीमांसका जगुः।
तस्याश्चेदं क्रियाङ्.गत्वमन्यद्वारकमात्मवत्।।

एवं षड् देया इत्यन्यं तद्वर्णमालभेतेति च तद्द्वारकमेव योज्यम्, एतेनोपचयापचयसंघातावपि वक्तव्यम्, सम्बन्धग्रहणादिकार्यं च जातिपक्षे एव सूपपादम्, आनन्त्यव्यभिचारादिचोद्यानवकाशात्।

यत्पुनरभ्यधायि प्रत्यक्षविषये पदं वर्त्तते न च प्रत्यक्षस्य सामान्यमात्रं विषय इति परिहृतं तद्वार्त्तिककृता प्रत्यक्षस्य हि चित्रात्मकं वस्तु विषयः।

न च तत्तादृशं किंचिच्छब्दः शक्नोति भाषितुम्।
सामान्यांशानपोद्धृत्य पदं सर्वं प्रवर्त्तते इति।।

न हि नानाधर्मनिचयरचितचित्राकारवस्तुसमर्पणनिपुणमेकं किमपि पद. मुपपद्यते।न च तत्र सम्बन्धग्रहणं सुकरमिति सामान्यांशनिष्ठमेव पदं युक्तं तस्माज्जातिरेव शब्दार्थ इति।

जातिमात्रवाच्यत्वाक्षेपः

अत्राभिधीयते, न जातिः पदस्यार्थो भवितुमर्हति। पदं हि विभक्त्यन्तो वर्णसमुदायो न प्रातिपदिकमात्रम्, तत्र च प्रकृतिप्रत्ययावितरेतरान्वितमर्थमभिधत्त इति स्थितम्, द्वितीयादिश्च विभक्तिः प्रातिपदिकादुच्चरन्ती प्रातिपादिकार्थत्वेन स्वार्थमाचष्टे युगपच्च त्रितयं विभक्त्यर्थः कारकं, लिङ्ग संख्या च न चैतत् त्रितयं प्रातिपदिकार्थे जातावन्वेति, न जातिः कारकं, न च जातेःस्त्रीपुंनपुंसकविभागः, न चास्या द्वित्वादियोग इति।

ननु व्यक्तिलक्षणया सर्वमुपपत्स्यते इत्युक्तम्, न च युक्तमुक्तम्, सकृत्प्रयुक्तं पदमंशेन कंचिदर्थमभिदधाति, ततोऽर्थान्तरं लक्षयति, तद्गतत्वेन पुनः लिङ्गसंख्याद्यभिधत्ते इति न प्रातीतिकोऽयं क्रमः।

साक्षात्तदन्वितत्वेन कथ्यमानं त्वसङ्गतम्।
तद्भवेदग्निना सिंचेदित्यादिविधिसन्निभम्।
ननु पुंसीव सामान्ये कथं नैव भविष्यति।
व्यक्त्यन्तरितमित्येषु युक्तो वैभक्तिकोऽन्वयः।।

प्रतिपदिकार्थसामान्यगतत्वेनैव विभक्त्या स्वार्थोऽभिधीयते न लक्षितव्यक्तिवृत्तित्वेन यतोऽभिधानवैषम्यं स्यात्स च विभक्त्यर्थो जातौ साक्षादसम्भवन्कक्ष्यान्तरितो भविष्यति, नैतत्, आत्मनो ज्ञानप्रयत्नादियोगित्वेन कारकत्वोपपत्तेः जातेस्तु साक्षान्न मनागपि व्यापारलेशसंस्पर्श इति स्वतो दुर्लभं तस्याः कारकत्वम्, अतः कथं तत्र विभक्त्यर्थान्वयः।

यदन्वितं च संख्यादि तत्स्थत्वेन न कथ्यते।
कथ्यते यद्गतत्वेन न तत्तेन समन्वितम्।।
न च व्यवहितव्यक्तिप्रतीतिं मन्यते जनः।
नाऽन्यथानुपपत्त्यापि क्रमसंवेदनं क्वचित्।।
जनयन्तीं च पश्यामो व्यक्तिं जात्यनुरंजिताम्।
संख्यादियोगिनी चेति सा वै धत्ते पदार्थताम्।।

जातेःकार्यान्वयः श्येनाचितमित्यभ्यधायि यत्।।

तत्र संस्थाननिर्देशङ्गन्न जातेः काचिदङ्गता ।।

श्येनव्यक्त्या चेत्सादृश्यमिष्टकाकूटस्य नास्ति व्यक्त्यन्तरेण व्यभिचारात् नितराममूर्त्तया जात्या सादृश्यमाकाशेनेव न तस्यावकल्पते इति स्पर्धोदाहरणमेतत् ।

जातिविशिष्टवाच्यत्वोपपादनम्

अन्येषु तु प्रयोगेषु गां देहीत्येवमादिषु ।
तद्वतोऽर्थक्रियायोगात्तस्यैवाहुः पदार्थताम् ।।
पदं तद्वन्तमेवार्थमांजस्येनाभिजल्पति
न च व्यवहिता बुद्धिर्न च भारस्य गौरवम् ।।
सामानाधिकरण्यादिव्यवहारोऽपि मुख्यः ।
वृत्त्योपपद्यमानःसन्नान्यथा योजयिष्यते ।।

तस्मात्तद्वानेव पदार्थः ।

ननु कोऽयं तद्वान्नाम तदस्यास्तीति तद्वानिति विशेष एव सामान्यवानुच्यते विशेषवाच्यत्वे चानन्त्यव्यभिचारौ तदवस्थौ सामान्यं तु शब्देनानुच्यमानं नोपलक्ष्यमाणं भवति उभयाभिधाने च शब्दस्यातिभार इत्युक्तम् ।

उच्यते, नेदन्तानिर्दिश्यमानशाबलेयादिविशेषस्तद्वान् न च सर्वस्त्रैलोक्यवर्त्ती व्यक्तिव्रातस्तद्वान् किं तु सामान्याश्रयः कश्चिदनुल्लिखितशाबलेया दिविशेषस्तद्वानित्युच्यते, सामान्याश्रयत्वाच्च नानन्त्यव्यभिचारयोस्तत्रावसरः,न च विशेषणमनभिधाय विशेष्यमभिदधाति शब्द इत्यभ्युपगच्छामः येनैनमतिभारेण पीडयेमहि, सामान्याश्रयमात्रे सङ्केतग्रणात् तावन्मात्रं वदतः शब्दस्य कतरोऽतिभारः एवं तद्वतोर्नास्वातन्त्रत्वादित्यादि दूषणं परिहृतं भवति ,किं च -

प्रत्यक्षं न हि निष्कृष्टजात्यंशपरिवेष्टितम् ।
तद्गोचरप्रवृत्तश्च शब्दस्तं कथयेत्कथम् ।।

तस्मात्प्रत्यक्षविषये प्रवर्त्तमानं तत्समानविषयमेव भवितुमर्हति पदं न सामान्यमात्रनिष्ठम् ।

युगपन्ननु संवित्तिर्विशेषणविशेष्ययोः।
प्रत्यक्षेऽपि न दृष्टैव न च युष्माभिरिष्यते।।
कार्यकारणभावो हि तद्धियोर्भवतां मते
तस्माद्विशेषणे जातौ पूर्वमिन्द्रियजा मतिः।।
पदादपि तदा यत्तु सम्बन्धज्ञप्त्यपेक्षिणः।
तत्रैव बुद्धिरित्येव न व्यक्तेरपि वाच्यता।।

उच्यते, प्रत्यक्षे तावद् द्वयोरपि विशेषणविशेष्ययोरिन्द्रियविषयत्वं सामान्येऽपि संयुक्तसमवायादिन्द्रियं प्रवर्त्तमानं विशेषणवद्विशेष्यमति विषयीकरोति, न हि सामान्यं प्रत्यक्षं विशेषोऽनुमेय इति व्यवहारः, एवं गुणमात्रग्राहिणीन्द्रिये गुणिनोऽनुमेयत्वं स्यान्न चैवमस्ति, तस्माद्विशेष्यपर्यन्तं प्रत्यक्षं तथा पदमपि तत्तुल्यविषयं न तु सामान्यमात्रनिष्ठमिति युक्तम्।

यत्तु "सामान्यांशानपोद्धृत्य पदं सर्वं प्रवर्त्तते" इति तत्केवलव्यक्त्यभिधाने सति आनन्त्यव्यभिचारभयादुच्यते तद्वदभिधाने तु तद्भयं नास्तीति न शुद्धजात्यभिधातृतया शब्दः संकोचनीयः, न तु सर्वात्मना प्रत्यक्षतुल्यविषयः शब्दः प्रतिपत्तिसाम्यप्रसङ्गात् न च शब्दादिन्द्रियाच्च तुल्ये प्रतिपत्ती भवतः, तदुक्तम् "अन्यथैवाग्निसम्बन्धाद् दाहं दग्धोऽभिमन्यते" इत्यादि।

उच्यते, पूर्वमेवैतत्परिहृतं सकलविशेषग्रहणाभ्यां प्रतिपत्तिविशेषसिद्धेः धर्म्यभिप्रायेण च संप्लवस्योक्तत्वान्नैतावता सामान्यमात्रनिष्ठः शब्दो भवति।

अपि च निष्कृष्टसामान्यांशवचनत्वे पदस्येष्यमाणे गोशब्दाद् गोत्वशब्दाच्च तुल्ये प्रतिपत्ती स्याताम्, गौः शुक्ल इतिवच्च गोत्वं शुक्लमिति बुद्धिः स्यात्, चातुर्वर्ण्यादिवच्च स्वार्थे एव गोशब्दाद्भावप्रत्ययस्त्वतलादिः स्यात्।

अथ मन्येथाः आक्षिप्तव्यक्तिकां जातिं गोशब्दो वक्ति भावप्रत्ययान्तस्तु निष्कृष्टत्वरूपमात्रनिष्ठामिति तदनुपपन्नम्, अनाक्षिप्तव्यक्तिकाया जातेः कदाचिदप्यदर्शनात्।

अथ गोशब्दश्रवणवेलायां व्यक्तिसंस्पर्शवती जातिरवगम्यते भावप्रत्ययान्ते तु गोशब्दे श्रुते तच्छून्याऽसौ प्रतीयते इति, यद्येवमागतोऽसि मदीयं पन्थानम् आश्रयवती चेज्जातिरुच्यते शब्देन आत्याश्रय उक्त एव भवति नान्यथा साश्रयवत्युक्ता स्यात्।

तदाश्रयपरिहारेणाश्रयिसामान्यमात्रविवक्षायां त्वतलादयः प्रयुज्यन्ते तथा चाहुः "यस्य गुणस्य हि भावाद् द्रव्ये शब्दनिवेशस्तदभिधाने त्वतलादय" इति, गुणस्य हि भावाद् द्रव्ये शब्दनिवेशः इति तद्वद्वाच्यत्वपक्षसाक्षीण्यक्षराणि, सामानाधिकरण्यं च तत्रैवोपपद्यते इत्युक्तमेतत्।

व्यक्तिबोधे पदव्यापारसाधनम्

तस्मात्-

यथा विध्यन्तपर्यन्तो वाक्यव्यापार इष्यते।
तथैव व्यक्तिपर्यन्तः पदव्यापार इष्यताम्।।

अनवरतव्यापारे शब्दे तदवगमात्।

येनान्विताभिधानं च पदानामभ्युपेयते।
सुतरां तेन वक्तव्या व्यवत्यन्ता पवतो गतिः।।
न हि व्यक्त्यनपेक्षाणां जातीनामितरेतरम्
अन्वयोऽनन्वितानां च नाभिधानमिति स्थितिः।।
गङ्गायां घोष इत्यादौ यथा सामीप्यलक्षणा।
नैव गौःशुल्क इत्यादौ गम्यते व्यक्तिलक्षणा।।
प्रयोगप्रतिपत्तिभ्यां वृद्धेभ्योऽध्यवसीयते।
तस्माद् गवादिशब्दानां तद्वानर्थ इति स्थितम्।।

तदिदमुक्तं सूत्रकृता "व्यक्त्याकृतिजातयस्तु पदार्थ" इति तुशब्दो विशेषणार्थः किं विशिष्यते गुणप्रधानभावस्य नियतेन शब्दार्थत्वम्, स्थितेऽपि तद्वतो वाच्यत्वे क्वचित्प्रयोगे जातेः प्राधान्यं व्यक्तेरङ्गभावः यथा गौर्न पदा स्प्रष्टव्येति सर्वगवीषु प्रतिषेधो गम्यते, कचिद्व्यक्तेः प्राधान्यं जातेरङ्गभावः यथा गां मुञ्चगां बधानेति नियतां कांचिद्व्यक्तिमुद्दिश्य प्रयुज्यते, क्वचिदाकृतेः प्राधान्यं व्यक्तेरङ्गभावो जातिर्नास्त्येव यथा पिष्टकमय्यो गावः क्रियन्तामिति सन्निवेशचिकीर्षया प्रयोग इति सर्वगतत्वेऽपि जातेर्न मृद्गवादौ वृत्तिरित्युक्तम्, तदेवं गवाश्वादिशब्दाना तावत्तद्वानर्थ इति सिद्धम्।

येषामर्थेषु सामान्यं न सम्भवति तैः पुनः ।
उच्यते केवला व्यक्तिराकाशादिपदैरिव ।।

* THE MEANING OF A WORD LIKE 'COW' ETC.

The Naiyāyikas hold that the judgement of perception is true. Since it relates to the external objects a person is never misled to identify the meaning. The Naiyāyikas, therefore, incline to discuss what actually a word refers to. It is consistent with their views that a word refers to an individual, its form and a universal.

Distinction Between the Meanings of *jāti* and *ākṛti* :

The term *ākṛti* refers to the form of an individual. It is just an organisation of limbs. Since it has been clearly distinguished in Nyāya Sūtra one should not follow the way of Jaimini who hold that *ākṛti* refers to universal.

Jaimini in his Mīmāṁsā sūtra admonishes that the term *ākṛti* refers to universal because it is useful in sacrifice. Śabara, the commentator on the Mīmāṁsā Sūtra sticks to the same interpretation and says that the arrangement of its limbs like dewlap etc. is useful in sacrifices. Kumarila has explained the point in his *ślokavārtika:*

The univeral is *ākṛti.* It determines the individual. It is a common property belonging to all individuals and is the source of the idea of oneness.

But the author if Nyāya sūtra distinguishes *ākṛti* from *jāti. Ākṛti* is just an organisation of parts. It is also consistent with common behaviour that *ākṛti* signifies an orderly arrangement of limbs of an individual. For instance, people say that 'merit reside in him who has a beautiful form'. Ākṛti, therefore, means an orderly arrangement of limbs.

*** This chapter has been taken from Nyāya-mañjarī, (āhnika 5.).**

Now, it is a matter of examination whether the form of an individual is referred to by a word or not. We should not neglect it like the followers of Jaimini. We should discuss who among the three namely individual, shape and universal constitutes the referent.

Individual as the Referent :

The persons who hold that the shape is the referent say that the meaning of a word is ascertained by two methods namely (i) the use of a word by a speaker and (ii) understanding of its meaning by a listener. An experienced person employs a word like 'cow'. What is grasped from the word stands as meaning of that word. If the word 'cow' is used to refer to an animal having dewlap but not having manes then the word signifies a particular form. The shape of an object is perceptible. It is also known through our sense organ that the shape of a cow is distinct from that of a horse. It is true that the shape of an object alone distinguishes the object itself from another. If the perceptible object constitutes the meaning of a word then the shape of an object should be its meaning. The shape of an object is indirectly related to action such as the act of sending etc. through individual.

But this does not stand to reason. Thus it is said :

Since the shape of an object varies in every individual and it is impossible to ascertain the re' :ion between word and the shape it would lead to apparent contradiction and infinte regress.

The word cow, therefore, does not refer to a particular form like *śābaleya* because it also refers to the shape of *bāhuleya* which is excusive of *śābaleya.* It is also imposssible to find out a common shape which belongs to all individual cows scattered over the three worlds and relate it to a word since the individual shapes are infinite in number. For this reason it is said that a word does not refer to shape. It is a fact that the shape is not related to action such as sending etc.

On hearing a sentence like 'Bring a cow' an intelligent person does not bring an object which bears the shape of a cow, viz. the picture of a cow or an earthen cow.

When we consider this from *Jātivādin's* standpoint the problem also remains unsolved. According to *Jātivādins* the universal is all pervasive. When a person says' "bring a cow" why does the listener not bring an earthen cow ? This is answered : Why does not the listener bring an elephant when the universal is all pervasive ? The Jātivādins say : Although the universal is present everywhere it is manifested through particular individual. But it is the case that the animal having dewlap etc. manifests the universal cowness. There is no incongruity in not bringing the earthen cow when one says 'Bring a cow'. Another point is that if a word refers to universal then the propositoon, 'A cow is white' becomes meaningless. Because the co-referentiality would not be possible among the words which refer to whiteness and objects etc. Moreover, the colour 'whiteness' does not belong to the universal of cowness. As a matter of fact since co-referentiality becomes predominant in relation to individual it is better to accept that a word refers to individual.

If individual constitutes the meaning of a word then a relation is to be established between individual and the act to be performed. If universal is the meaning the relation is easily grasped. But shape as the meaning is neither related to action nor to a word. How does *ākṛti* constitute the meaning ? Now we should determine which of two namely a universal or an individual constitutes the meaning of a word. The persons who hold that a word refers to individual say to this effect.

The word refers to individual because there is a consistency of the meaning of a word with injunctions governing the performance of the vedic rites. If universal is referred to by a word then the injnuctions relating to cutting, killing and sprinkling etc. become incongruous since nobody can cut, or kill or sprinkle a universal. Moreover, there are injunctions which enjoin thus: "One should make a gift of 6 articles, of 12 articles, of 24 articles etc." In these cases the number six, etc. cannot be connected with a universal. It is individual therefore, referred to by a word. It is also a point that in a sacrifice if an animal is offered to a deity but it manages to escape then an injuction runs to the effect another animal of the same class having the same clolur and age should be slain. If universal is referred to in such a situation then killing of the animal would be highly impossible.

Another animal which has been brought now possesses one and the same universal. Hence, it cannot be qualified by the 'another'. The adjective another is applicable to individual only. Hence, the individual is referred to by a word.

If individual is referred to by a word then bringing together of animals, the increase of number, their reduction of number, the killing of them and the relation of ownership etc. are compatibly explained.

If individual is secondarily referred to then it does not fit in things. If the straight path is available one should not go by a curved one.

Although a word refers to an individual yet it has also secondary power of expressing universal. Thus there would be no possibility of assuming infinite number of relations and that of misrepresentation.

Moreover, some Logicians hold that a word refers to perceptible objects. A universal is not the object which is perceived. You should therefore, admit that individual is only referred to by a word. Whenever a word is heard the individual is only known.

UNIVERSAL AS THE REFERENT

The followers of Jaimini do not subscribe to this view. They say : Whether the referent is mere individual or individual characterized by universal ? Mere individual cannot be the referent. Because no speaker employs a word to refer to mere individual and no listener understands a particular cow. It may be said that the word cow refers to individual cow characterized by universal cowness. If it is cowness then the individual is not the referent. Why ? If particular individual constitutes the referent then we cannot employ the same word to refer to another individual. If a word is employed to refer to another individual then we have to admit that the referent is a common property shared by all individuals but not as particular individual.

The Vyaktivādins defend themselves that 'another individual' is individual only. The word cow is employed to refer to such an individual but not to universal. But this does not hold good. If it refers to individual then it should refer to particular white horse since it is an individual. The Vyaktivādins argue that wherever we find such type of use we employ the word to refer to so. Nobody has used the word 'cow' to refer to a new born calf since it has not been found to be employed in that sense. For this reason, our acquaintance with usage of words does not justify the stand that a word refers to an individual. It is impossible for a word to refer to each and every individual. If it is accepted that a word refers to only a particular individual then on hearing 'cow' one should determine that this is only a particular cow but not that this is also a cow. One cannot employ a word at his discretion if there would be nothing to regulate its meaning. There should be a regulating factor. Now it may be said that the universal 'cowness' regulates the use of the word cow. But whether this universal is known or unknown ? If it is unknown then it should be held that from which source it is known. Whether it is from a word or from other source. It is not from other source since there is no proof to this effect. If cowness is known from the word then it should be pointed out that the word 'cow' is at the outset related to the universal 'cowness' since there is a rule that a qulified object is not at all known if the qualification is not grasped.

If the word 'cow' refers to individual as a qualificand and to universal as a qualifier then it is heavy for a word to be expressive of such things. But there is no other alternative for presentation of the individual. In fact, we donot conceal the awarness of an individual. We also donot conceal the awareness of universal. Because both these are presented to our consciousness. The power of conveying both individual and universal is too heavy a load for a word. The awareness of any one of them points to that of another. If it is the case then whether the word cow referring to universal indicate the individual or it signifying individual points to the universal ? As such, universal becomes the meaning of a word since the universal qualifying an individual is firstly presented to our consciousness. When universal is conveyed the awarness of an individual follows from it. Thus the hypothesis that the word has an expressive power to refer to both individual and universal is not acceptable.

As we know both qualifier and qualificand from the word *daṇḍī* 'staff-holder' so from the word cow we can understand that the individual cow is a noun and universal is a qualifier. How far it carries the heavy load ? But this has no bearing upon the point in question. These two aspects are not known from division of stem and suffix. The word *daṇḍī* is derived from a stem with a suffix *in*. The stem *daṇḍa* points to the qualifier. The suffix has a sense of possession. It refers to the qualificand. But this logic is not applicable to the word cow. In the above example the word *daṇḍī* does not point to the qualifier and the word *daṇḍa* does not refer to the qualificand. But in case of the word 'cow' it is understood that it is non-composite word which consists of no parts. It stands for either a qualificand or a qualifier. If it denotes a qualificand and has the sense of a qualifier then it does so on receiving the aid of some other sources of knowledge. If it denotes only qualifier then it may point to qualificand because the qualifier cannot stand unsupported. It is our perception that when a word is uttered the individual is presented to our consciousness. But there is a doubt whether this is communicated from the word itself or from the universal. The distinction between these two cases of knowledge is not perceived by us. The problem is to be solved by means of argument. It is difficult to prove that a word simultaneously conveys two meanings namely qualifier and qualificand. It cannot be said that it successively conveys two meanings viz. it conveys the first meaning at the outset, stops for sometime and then conveys the second meaning. If there is no word, the universal is understood and from the knowledge of universal the individual is known. This shows that the individual is known from the universal itself. But the universal is known from the word. The Naiyāyikas do also admit that a judgement presupposes the awareness of a predicate as its antecedent condition. Kaṇāda has also stated that "The colour 'whiteness' inheres in a substance. When we know that the said colour is in a substance we know that the substance is white. The awareness of the white colour inhering in a substance and the judgement that the substance is white are causally connected".

It has been said that since the verbs such as to cut, to kill, to sprinkle etc. are related to individuals the individual stands as the referent of a word. This is also fallacious because in some cases the universal is seen to be connected with verbs. For instance ***śyenacitaṁ cinvīta*** 'one should prepare fire in the shape of hawk.'

In this case the individual also subservient to a vedic rite. i.e. an action, since a universal, a formless being, cannot be serviceable. Such a contention does not hold good here. In the above rite a hawk is not mentioned as instrumental to it just as an animal is referred to as instrumental to a vedic rite in the sentence ***paśunā yajeta.*** The designation of a vedic rite as fire is the title of this section. The wording this text purports to convey that a hawk should be constructed after observing few rites. A bird, hawk, cannot be created by observing a few rites. We also remember what has been stated in the chapter devoted to the naming of fire. It has been stated that a hawk cannot be created by piling bricks. But fire is so prepared that it assumes the forms of a hawk. Hence a universal is subservient to a rite but not an individual. It cannot be said that the resemblance of fire to a hawk will be constructed. Because such a contention does not hold good since resemblance is not a thing to be done. What is similar to a particular hawk is also dissimilar to another hawk. Hence, dissimilarity accompanies resembleance. The objection that a universal being formless, is not subsevient to a rite is baseless. Although an attribute and an action are formless yet their efficacy in connection with the performance of some rites is acknowledged. For instance, one purchases a red cow and one offers libation into fire going around it'. Here we see that a universal is also indirectly related to an action since an individual is implied by a universal. Therefore, the verbs such as 'to cut', 'to kill' and 'to sprinkle' etc. are indirectly related to a universal.

A rite which is practicable by means of an individual implied by universal is said to be practicable by the universal itself just as an action which is performed by the body and sense-organs is asserted to be done by the soul.

The soul is enjoyer of the fruits of all actions. It is the doer of all actions. Since it is formless it gets everything done by means of body and sense-organs. It cleans the twigs of a figtree, observes the sacrificial ghee and does such other things. Thus becomes the agent of an action, i.e. doer. Similarly, a universal performs an action through the help of an individual and becomes instrumental to an action, through the help of an individual and becomes instrumental to an action.

The Mīmāṁsakas, therefore, hold that a universal is instrumental to an action. But it performs its action through the activity of something else just as the soul becomes the doer of an action.

In this light the example 'one should make six gifts' may be examined. If an animal dedicated to a God in a sacrifice runs away then another animal bearing a close resemblance to it should be killed. In this case a universal is instrumental to these rites through an individual. Following the same line of thinking growth, reduction and aggregation etc. may be applied to universal. If it is held that the universal is the referent then the relation between a word and its referent is easily cognised. The defects such as infiniteness and contradiction etc. have no scope in this case.

It has been also objected that a word denotes a perceptible object and a universal is not the exclusive object of perception. This objection has been met by Kumārila who holds that a concrete object having various properties is the object of our perception.

Kumārila says that there is not a single word which is logically conceived as competent to denote an object having various properties. Thus there is a restriction of power of denotation. A word, as such, refers to universal which is one of the various properties of an object. The universal thus becomes the referent of a word.

Refutation of Universal as the Referent.

It is said here that a universal cannot be the referent of a word. Because a *pada* 'word' is the combination of letters with case-endings. But it is not a mere stem. It is the principle that the stem and the suffix express a mutually related meaning. When a suffix such as the second case-ending etc. is attached to a stem the meaning of an inflection is presented to our consciousness derived from the base. The meaning of inflection is three fold, namely, case, gender and number. But these three cannot be related to universal. A universal cannot be a case. It has got no gender like masculine, feminine or neuter. Moreover, it cannot be marked with a number such as 'two' and 'three' etc.

It is contended that all these difficulties will be solved by way of making the individual referent secondarily. But this is not satisfactory. A word is once employed and expresses certain meaning partly by its primary import. Then the another meaning is implied secondarily. It thus expresses case, number and gender. But this order of presentation of meaning in its relational character is not borne out by our experience. It is improper to hold that the universal is directly related to inflection. It is just like a contradictory injunction that one should sprinkle with fire. Why is it not tenable to hold that as the meaning of an inflection is related to male so it is related to universal through the individual ?

An inflection conveys its meaning being related to the meaning of a stem, i.e. the universal, but not being related to the secondary meaning i.e. individual. If it does so then there will be no uniformity in conveying the primary meaning of a word. The Mīmāṁsakas hold that as it is impossible to establish a relation between the primary meaning of a word and that of an inflection so the meaning of an iflection is related to an individual. Such a defence is not tenable. A soul is logically called the author of an action since it has consciousness, volition etc. But a universal is no way related to any action. So there is no possibility for universal to stand in case-relationship. How will then the meaning of an inflection be related to a universal ?

The number such as 'one' and 'two' etc. is said to be syntactically related to a universal. In fact it is not related to universal. But it is said that the number belongs to such a substratuṁ with which it is not syntactically connected.

Nobody experiences that a number is indirectly related to a universal through individual. If a number is not related to an individual at first then it is logically asburd for it to be related to a unviersal. Such a gradual development of knowledge actuated by the necessity of removing inconsistency is not experienced.

We see that an individual which is the generator of an action is marked off by universal and is qualified by a number etc. is expressed by a word.

The Mīmāṁsakas hold that a universal is directly related to a Vedic action. They cite an example like 'one should arrange fire in the shape of a hawk' for this purpose. In this case there is no reference to universal. Hence a universal is not directly subservient to an action.

The Mīmāṁsakas have said that the pile of bricks containing fire does not bear a resemblance to a particular hawk since it is similar to one hawk then it may be as well dissimilar to another hawk. The argument is not convincing because the universal of hawkness is formless and it is highly inconsistent to hold that the said pile is similar to universal. It is as good as to say that it is similar to the sky. The example simply represents an evading spirit.

Presentation of Meaning Qualified by Universal

There are many other usages like 'Give a cow' etc. in which an individual having the universal is syntactically connected with the verb. So it is held that the referent is individual possessing the universal.

A word primarily refers to one characterised by universal. It is not sound to hold that the meaning of inflection is at first related to individual and then becomes related to universal. This view is not tenable.

If it is said that the primary meaning of a word is an individual characterized by a universal then the use of cases in apposition finds justification. If we donot subscribe to this view then the co-referentiality subsisting between a subject and a predicate cannot be explained.

A word, therefore refers to the possessor of a universal.

The Mīmāṁsakas raise objection. What is the meaning of *tadvān* 'the possessor'? The literal meaning is that 'this has that'. The demonstrative pronoun 'this' stands for a near individual. Hence the meaning of the word amounts to an individual having a universal. If it refers to particular then it is open to the same charge that a word has an infinite power of conveying meaning and it refers to objects which fall outside the scope of its meaning. Since there is no word to refer to universal so a universal cannot be used as an objective. If it is held that a word refers to both individual and universal then it would be heavy burden for a word. This has already been stated.

The Naiyāyikas meet this objection in the following way. The demonstrative pronoun 'this' (in *tadvān*) does not stand for a distinct particular such as *śābaleya*. It also does not refer to all individual cows in the universe. But it stands for any individual cow, the substratum of the universal but does not refer to a distinct individual cow such as *śābaleya*. Since it resorts to universal the above charges that a word has infinite power of referring to objects and has unjustifiably wide denotation become pointless. We do not admit the position that a word does not refer to noun without referring to its qualifier. So we do not overburden word with the weight of its meaning. As we simply grasp the relation between a word and its meaning just resorting to universal so how much is the weight of meaning on the part of a word? The charge that if a universal is not distinct from its substratum then a word cannot refer to both an individual and a universal does not affect our position since all these objections have already been met with.

Moreover, perception does not reveal a universal only. A universal is a property of its substratum i.e. individual. Perception reveals both. When a word is employed to refer to an object of perception how is it that it refers to only a universal ?

Thus, we arrive at the conclusion that if a word is employed to refer to an object as perceived then a word should also mean the same object. There should be perfect co-ordination between the meaning of a word and the content of an act of perception. As a matter of fact, a word does not rest on universal only.

The Mīmāṁsakas raise objection that qualifier and qualificand are not simultaneously perceived. You (the Naiyāyikas) do not admit such a perception. The direct awareness of qualifier and qualificand is casually related to each other. Hence, the sense perception of a universal takes place at first. Thus a word has a reference to a universal in point of the expressive relation. The universal stands as the referent and there is no scope for individual.

The Naiyāyikas reply that both qualifier and qualificand are perceived. Whenever a universal is perceived the sense-object contact in the shape of conjunction-cum-inherence is required. For this reason whenever a sense-organ senses a universal it senses an individual, the substratum of universal. We donot say that a universal is only percieved and thc qualificand i.e. the individual is inferred. If it is admitted then we should admit that our sense organ senses an attribute but the substance which possesses it is inferred. But there is no such provision. We therefore, hold that substratum of a universal is also perceived. As there is a co-ordination between meaning and the content of perception so it is not reasonable to hold that a word refers to universal.

Kumārila has stated that a word is employed to refer to the elements of universal. If we hold that a word refers to only an individual then we have to admit that a word has expressive power to refer to innumerable individuals and it would transcend its scope of expressive power. There is no difficulty if we say that a word refers to the possessor of universal. We should not, therefore, impose limitation upon the meaning of a word, subscribing to the view that a word merely refers to a universal. It is also not the case that the perception of a word is identical with that derived from sense organ. Infact two types of knowledge are not identical. Thus it is said to this effect 'A man who has been burnt differently feels the actual contact of fire with his body'. In other words, the knowledge of meaning of the word 'burning' is entirely different from the actual experience of burning.

There is no question of *pramāṇasamplava* as it has already been discussed and discarded. Although we admit that there is a difference existing between perceptual and verbal knowledge since the former reveals all the details of the specific character of an object and the latter does not do so. Yet these two types of knowledge refer to the same object as they refer to the same substratum. Even if we refute the *pramāṇasamplava*, it does not follow that a word refers to only universal.

Moreover, if it is admitted that a word refers to only a universal then word 'cow' and 'cowness' should refer to the same thing. There should be the judgement that cowness is white like the judgement 'A cow is white'. The suffixes *tva* and *tal* etc. conveying an abstract property should only signify the meaning of the stem to which they are attached like the word *cāturvarṇya.*

If the Mīmāṁsakas think that the word 'cow' refers to the universal cowness but implies an individual then it is contrary to reason to hold that the word 'cowness' which has been derived from the word *go* 'cow' with *tva* suffix conveying an abstract property refers to universal. A universal without indicating individual is not found.

If you say that when the word 'cow' is heard the universal 'cowness' is known as associated with individual. But when the word 'cowness' is heard a universal devoid of individual is known. If it is held that a universal having a locus to stand upon is referred to by a word then it amounts to this that a word refers to the substratum of a universal, otherwise, the universal having the substratum cannot be asserted. When we intend to express only a universal as independent of its locus we use words ending in suffixes like *tva* and *tal* etc. Patañjali in his *mahābhāṣya* has said that "as the concrete objects are referred to by nouns so nominal suffixes like *tva* and *tal* etc. are attached to these nouns. This remark of Patañjali substantiates the stand that a word refers to concrete individual. Thus, the proposition 'A cow is white' is squarely explained in point of its coreferentiality between the subject and the predicate.

Establishment of the Function of a Word to Refer to an Individual.

The Mīmāṁsakas admit that a sentence, in the subjective mood continues to convey its meaning until and unless all the minor details which are necessary for the principal act are expressed. Similary, he has to admit that a word does not completely convey its meaning until and unless it expresses an individual. As a word refers to an individual by uninterrupted process of conveying its meaning so an individual is also included in its primary meaning.

The Mīmāṁsakas hold that words in a sentence always reveal an inter-related meaning. A word, therefore, reaches the culminating point when it refers to an individual. Further, the universals donot get themselves inter-related without a reference to individual. The Mīmāṁsakas cannot also hold that disconnected universals are referred to by words since they say that words never present disconnected objects as their meanings. As the word *gaṅgāyaṁ* 'in the *Gaṅgā,* in the sentence *gaṅgāyaṁ ghoṣaḥ'* there is a village of cowheards on the banks of the river *Gaṅgā*, indicates a place, adjacent to the flowing river, on the strength of its secondary meaning so the word 'cow' in the sentence 'A cow is white' does not mean as individual cow by means of its secondary meaning. The meaning of a word is learnt if we carefully understand the meaning of said word noting the usage of experienced persons. Thus it amounts to saying that the word like cow refers to the possessor of universal.

The author of Nyāya-Sūtra has said that the meaning of a word consists of an individual, a universal and a shape. The particle *tu* in the *sūtra* has been used to determine something. What does it determine ? It determines the primacy of a particular meaning according to merit of the same. Although the individual constitutes the meaning of a word in some cases the universal becomes prominent and the individual occupies a subordinate position. For instance, 'A cow should not be touched with foot'. Since it is a universal prohibition it applies to all cows. In some cases the individual plays an important role and the universal becomes subservient. For instance, 'Untie a cow' and "Enchain a cow" etc. These expressions are used with reference to the individual. In other cases, the shape become predominant, individual is subservient and there is no universal. For instance, 'Prepare cows from powdered rice'. If we use sentences for making a shape then the universal does not belong to earthen cow and similar objects although the universal is all pervasive. The absence of universal in such objects has already been established. The meaning of a word, therefore, is the possessor of universal i.e. the individual.

There is an impossiblility for few words to refer to universal. They refer to individual only. Words like sky etc. constitute examples of such cases.

* वाच्यार्थवैविध्यम्

एवं डित्थादिशब्दानां संज्ञात्वविदितात्मनाम् ।
अभिधेयस्य सामान्यशून्यत्वाद् व्यक्तिवाचिता ।।

अत एव हि द्रव्यशब्द इत्युच्यते।

ये पुनःकल्पितानेकभेदवृत्तिं प्रचक्षते ।
वाच्यं तत्रापि सामान्यमतीव ग्राहिकास्तु ते ।।
न हि डित्थत्वसामान्यं दृश्यते गगनत्ववत् ।
कल्पनायास्तु नो भूमिः काचिदस्ति विपश्चिताम् ।।

गुणशब्दास्तु केचित् स्वजात्यवच्छिन्नं गुणमभिधाय तावत्येव विरिमन्ति केचिद्गुणमभिधाय द्रव्यमाक्षिपन्ति तत्सामानाधिकरण्यप्रयोगदर्शनात् ।

गुणैकनियतास्तावद्गन्धरूपरसादयः ।
गन्धत्वादिव्यवच्छिन्नगन्धादिगुणवाचिनः ।।
तेषां न द्रव्यपर्यन्ता वृत्तिःक्वन दृश्यते ।
न गन्धः पद्म इत्यस्ति सामानाधिकरण्यधीः ।

न ह्येवं केचन वक्तारो भवन्ति ओदनं गन्धः, आम्रं रस इति, गुणं शुक्लादिशब्दास्तु कथयन्तस्तदाश्रयं द्रव्यमप्याक्षिपन्त्येव शुक्ले शुक्ल इति दर्शनात्,शुक्लादयोऽपि शब्दा गुणपदोपबन्धसङ्कोचितशक्तयों भावप्रत्ययान्तवद्गुणमात्राभधाने एव पर्यवल्क्यन्ति शुक्लो गुणोऽश्वे शोस्यमश्वे इति सामानाधिकरण्यप्रयोगदर्शनात् ।

क्रियाशब्दाश्च द्विविधा भवन्ति केचित्कर्त्तरि कर्मणि करणें वा प्रयुज्यन्ते केचिद्भावमात्रवचना एव, कर्त्रादिवाचिनस्तावन्निमित्तीकृत्य कांचन क्रियां तद्योगिनि द्रव्ये वर्त्तन्ते पाचकादयः।

---

* न्यायमञ्जरीत उद्धृतमिदं प्रकरणम् (न्या. म. ५)

यत्रापि तत्क्रियायोगस्तदानीं नोपलभ्यते।
तत्रापि योग्यतां दृष्ट्वा शब्दं तज्ज्ञाः प्रयुंज्जते।।

न हि पाचको लावक इति प्रवृत्तक्रिय एवोच्यते अन्यदाऽपि तथा-तथा व्यवहारात्।

अन्ये पूर्वापरीभूतस्वभावपरिहारतः।
सिद्धरूपतया प्राहुः शब्दाः पाकादयः क्रियाम्।।

तदुक्तम्, कृदभिहितो भावो द्रव्यवद्भवति क्रियावच्चेति, न चैकान्तेन परिहृतसाध्यमानावस्थैव तैः क्रियोच्यते यत आह - क्रियावच्चेति।

अन्यत्प्रवृत्तौ शब्दस्य निमित्तमवगम्यते।
अभिधेयं तु तस्यान्यदित्ययं प्रथमःक्रमः।।

पाचकादिशब्दानां हि प्रवृत्तिनिमित्तं क्रिया अभिधेयास्तु कर्त्रादयः-

क्वचित्पुनर्यदेवास्य स्यात्प्रवृत्तिनिबन्धनम्।
तस्यैव वाच्यता भावप्रत्ययान्तपदेष्विव।
इत्येवं लेशतस्तावन्नाम्नां वृत्तिरुदाहृता।
आख्यातानां तु वाच्योऽर्थःपुरस्ताच्चर्चयिष्यते।।
उपसर्गनिपातानां नाम्नामप्यविभागतः।
प्रयोगप्रतिपत्तिभ्यामनेकार्थोऽवगम्यते।।

उपसर्गः प्राचुर्येण क्रियायोगे प्रवर्तते, "उपसर्गःक्रियायोगे" इति स्मरणात्, केचितु नामभिरपि सम्बन्ध्यन्ते यथा प्रगतं वयो यस्य स प्रवया इति। अन्ये धातोविचित्रार्थतामापादयन्तस्तद्वशषणतया सम्बन्ध्यन्ते।

उपसर्गवशाद्धातुरर्थान्तरविलासकृत्।
विहाराहारसंहारप्रहारपरिहारवत्।।

ननु क्वचिदुपसर्गो धात्वर्थमेव बाधमानो दृश्यते यथा प्रस्थित इति गतिनिवृत्तवाची धातु प्रत्युत गतिवचनतां नीतः प्रशब्देन, न चेदृशं विशेषणं भवितुमर्हति।

येन स्वार्थाविरोधेन विशेष उपजन्यते ।
विशेषणं तदेवेष्टं न तु यत् स्वार्थनाशनम् ।। इति ।

नैष दोषः, अर्थान्तराभिधानसामर्थ्यापादनमेव धातोर्विदधदुपसर्गो विशेषणं भवितुमर्हति, अनेकार्थाभिधानशक्तिश्च धातुरुपसर्गेण नियतेऽर्थेऽवस्थाप्यते इति तस्य तद्विशेषणता ।

उपसर्गाः किमर्थस्य वाचका द्योतका इति ।
प्रकृतानुिपयोगित्वादिहैतन्न विचार्यते ।।
अन्वयव्यतिरेकाभ्यां तदर्थो ह्यवधार्यते ।।
तदागमे तत्प्रतीतेस्तदभावे तदग्रहात् ।।

ते तु किं वाचकाः सर्वे तदवगतिमुपदधति किं वा द्योतका इति किमनेन, एवं समुच्चयादिवाचिनां चादिनिपातानां "वृक्षं प्रति द्योतते" इति कर्मप्रवचनीयानामर्थः प्रयोगप्रतिपत्तिभ्यामवधारणीय इत्यलं प्रङ्गेन ।

अयमस्य पदस्यार्थ इति केचित् स तेन वा ।
योऽर्थः प्रतीयते यस्मात्स तस्यार्थ इति स्मृतिः ।।
स्पष्टामपि तु ये बुद्धिं निरूपयितुमक्षमाः ।
तां दर्शयितुमस्माभिर्दिङ्.मात्रमुपदर्शितम् ।।
इति प्रमाणत्वसमर्थनाय शब्दस्य किंचिद्वयमुक्तवन्तः ।
पदाभिधेयार्थनिरूपणं तु शास्त्रान्तरे विस्तरतःप्रणीतम् ।।
तत्क्षोदेन न नः प्रयोजनमतिद्रघीयसी सा हि भू-
रन्यात्मेव दिशं परीक्षितुमतोऽस्माभिर्गृहीतः क्षणः ।
जात्याद्यार्थसमर्थनेन कथिता बाह्यार्थसंस्पर्शिता
शब्दानामियतैव नन्ववसरेऽमुष्मिन् कृतार्था वयम् ।।

* MANIFOLDNESS OF MEANING

Thus there are words like *Ḍittha* etc. which are well known as names. Since they are devoid of universals they refer to individuals. It is, therefore, called as *dravya-śabda* 'the name of substance.'

Those who hold that a universal belongs to individual which admits of imaginary distinctions should also hold that a universal resides in the individual like *Ḍittha* since there are several individuals designated by the same name. As there exists no universal as skyness so there exists no universal as *Ḍitthatva*. Where is the scope for the wise to indulge in pure imagination?

There are words which refer to qualities being characterised by the same. There are some other words which having referred to qualities imply substances secondarily. This is so because they are found to be used in coreferential setup.

Words like scent, sound, teste etc. signify only qualities which exclusively belong to them as the specific characteristic features. They do not imply substances. The proposition that lotus is scent is never used to mean that scent is predicated of a lotus in coreferential situation. Nobody says that rice is scent and mango is taste. There are some words such as *śukla* 'white' which refer to qualities but imply substances. It is seen that the word *white* is used to convey a white substance. But when the word *white* is associated with *quality* it does not imply a substance as its power of implying a substance is restricted. These words

* **This chapter has been taken from Nyāya-mañjarī, (āhnika 5.).**

having suffixes indicative of abstract qualities whiteness etc. exclusively denote qualities. A horse has white colour. Whiteness is in a hourse. White colour and whiteness refer to one and the same thing when used in coreferential situation.

There are two kinds of verbal words. Among verbal words some denote nominative case, some objective case, some instrumental case, some other cases. Some verbal words exclusively denote verbs. Such words refer to verb at first, and then signify a substance which is connected with it. For instance the word *pācaka* 'cook'.

Where the connection with verb is not directly noticed the experts use such words ascertaining their compatibility.

When the words like *pācaka* 'cook' and *lāvaka*' 'cutter' etc. are used one does not mean to say that a cook is cooking and a cutter is cutting. It is marked that a person is called a cook or a cutter even when he does not cook or he does not cut.

A verbal word does not refer to the process of being or becoming. But it refers to an accomplised fact. There is also no reference to the different stages of an act which precede and succeed. The verbal word which is derived from a root with a suffix has a double character. It behaves both as a substance and a verb. When it represents the verbal character it does not mean that everything has been done and nothing remains to be done. So it has been said that it looks like a verb.

Another way of understanding a word is that its use and its primary meaning are not one and the same thing. This is the first step to ascertain the meaning. For instance, the word *pācaka* 'cook' the ground for its application is the action. The nominative case of such an act of cooking is its primary meaning.

There are few words which are derived from a root with a *kṛt* suffix attached to it. The ground for application of such words is identical with their primary meaning. Thus we have stated the primary meanings of nouns. We shall discuss the meaning of verbs later on. It is known from usages and experiences that prepositions indeclinables, and nouns admit of various meanings. In most cases, prepositions are prefixed to verbs. There is a general rule that some indeclinables become prepositions when they are prefixed to verbs. Some prepositions are annexed to nouns, viz., *pravayāḥ* 'one advanced in age'. Other prepositions modify verbs and determine their meanings in various ways. Thus has been said to this effect.

A root admits of various meanings when it is prefixed by various prepositions. If the root $\sqrt{hṛ}$ is prefixed by the preposition *vi* then it amounts to the root 'to play' (*vihāra*). When ā is prefixed to it the root becomes 'to eat '(*āhāra*). if it is preceded by *sam* it is identical with the root 'to kill (*saṁhāra*). When *pra* is prefixed it signifies the act of beating (*prahāra*). If it is preceded by *pari*, it is synonymous with the root 'to give up' (*parihāra*).

Sometimes, the prefix cancels the meaning of a root. For instances, the word *prasthita*. The meaning of the root $\sqrt{sthā}$ is 'to stop.' When it is prfixed by *pra* it expresses the sense of going. In fact, an adverb should not contradict the original meaning of the root to which it is expected only to modify.

Our view is that an adjective which characterizes a noun but does not mar its sense is a true adjective. An adjective does not repudiate the meaning of its noun.

Whenever a preposition is connected with a verb it brings about a change in the meaning of the original root. Thus doing so it becomes adverb. Although a root admits of a power of conveying various meanings being associated with a preposition yet it refers to a definite meaning. The function of a preposition is that it channalises the meaning of a root. For this reason a preposition is an adverb.

The point that a preposition either signifies the meaning or indicates it is not brought here for discussion because it is irrelevant to the issue. The meaning is ascertained by the joint method of agreement and difference. If the preposition precedes the root then it conveys certain meaning. If it does not precede then it does not convey the same.

We do not want to enter into the discussion whether a preposition conveys or indicates a meaning. There are a few indeclinables such as *ca* etc. which express the sense of conjunction. There are a few prepositions which are used independently of verbs. For instance, *vṛkṣaṁ parti dyotate.* The meaning of such prepositions will be determined by experience and usages. This topic needs no further illustration.

Some others hold that this word is expressive of this meaning, or, if a particular meaning is conveyed by a particular word, that meaning is said to be expressed by that word. We have shown here a direction for those who have power of understanding but cannot make out the meaning of a word. We have said something to establish the validity of verbal knowledge. The nature of a word and its meaning has been elaborately discussed in other works. We feel no necessity of discussing the same topic over and again since the subject matter is highly extensive. Now we have found leisure of discussing other topics. When we have supported the universal etc. as the meaning we have said that words are related to external objects. In the mean time with this discussion on words we are really satisfied.

* सार्थकशब्दनिरूपणम्

शब्दान्तरमपेक्ष्यैव सार्थकः स्वार्थवोधकृत् ।

प्रकृतिः प्रत्ययश्चैव निपातश्चेति स त्रिधा ।।

यादृशः शब्दः शब्दान्तरं सहकृत्यैव स्वस्य स्वघटकस्य वा वृत्त्युपस्थाप्ययादृशार्थावगाहिबोधं प्रत्यनुकूलः,स तथाविधार्थे सार्थकः। पटपाचकाद्याः प्रकृतयः सुप्तिङ्.ङ्ाद्याः प्रत्ययाश्चादयो निपाताश्च स्वोपस्थाप्यार्थस्य बोधं नियमतः शब्दान्तरं सहकृत्य जनयन्ति, वाक्यानि पुनरसहकृत्यापि, शशविषाणादिकः शब्दोऽपि शशीयत्वादिना विषाणादेरन्वयवोधमादधान-स्तादृशविषयताकबोधने सार्थक एव, परन्त्वयोग्यः।

स चायं त्रिविध एव - प्रकृतिः प्रत्ययो निपातश्चेति। शप्शयनादिविकरणो नुमाद्यागमश्चाऽन्वयबोधने प्रकृत्यादीनां सहायोऽपि न स्वार्थोपस्थितिद्वारा, येन तमादाय चतुरादिविधः सार्थकः स्यात्।"पुत्रो व" इत्यादावपि युष्मदादिसम्बन्धो न वसाद्यादेशस्यार्थः, किन्तु तदादेशिनो युष्माकमित्यादिवाक्यस्य, अतः पूर्वपदमपेक्ष्यैव तादृशसम्बन्धबोधानुकूले तत्र न प्रसङग्.ः। सदृशाद्यर्थे प्रकृत्यादिभ्यो भिन्नेषु निभनीकाशादिशब्देष्वतिप्रसङग्.स्तु समासान्तर्गतत्वनियतस्वार्थबोधकताकान्यत्वेन विशेषणाद्वारणीयः।।६।।

---

* शब्दशक्तिप्रकाशिकात उद्धृतमिदं प्रकरणम् (कारिका - ६)

* VALID WORD

The word is valid in point of its meaningfulness, when it is associated with the word used in juxtaposition. Such words are divided into three categories, viz, *prakṛti* 'Stem' *pratyaya* 'Suffix' and *nipāta* 'Particle'.

The word which presents its individual meaning or implicatory meaning in association with other words through the process of verbal relation and is conducive to the cognition of the same, is to be treated as valid in point, of its meaningfulnesss. In this point, the instances are the stems like cloth and cook etc, the case terminations and the verbal terminations, and the particles which present their associative meaning regularly. Sometimes, words present their meaning without being associated. For instance, the word *śaśa-viṣāṇa* 'hare's horn' has certain expectancy to accommodate the other constituent *viṣāṇa* 'horn' for a cognitive agreement. But it is ontologically incompatible.

Such word is of three kinds : stem, suffix and particles. Although the conjugational transformations like *śap* and *śyan* etc. and the augments like *num* etc. help the stem to bear out the knowledge of what the stem actually expresses in association with suffixes, they are not to be treated as valid words for the reason that they don't have their own meaning.

Similarly, in the expression *putro vaḥ* 'your son' the term *vas* as the substitute of *yuṣmākam* does not point to the relation of *Yuṣmad*. On the other hand, it presents the

---

*** This chapter has been taken from śabdaśaktiprakāśikā. V. 6**

meaning of substituted term *yuṣmākam* 'your' which is a sentence itself. The context of associative cognition i.e. the word produces cognition in association with other words) therefore, does not arise here.

The words like *nibha* 'similar' and *nīkāśa* on the same way, although point to the meaning 'similarity' are not to be taken as valid. Such words are meaningful being the constituent of a compound and not otherwise, i.e. such words are singled out by the adjective 'being constituent of a compound in which they refer to their own meaning and not otherwise.'

"नाम निरूणम्"

1- निरुक्ता प्रकृतिर्द्वेधा नाम- धातुप्रभेदतः।
यत् प्रातिपदिकं प्रोक्तं तन्नाम्नो नातिरिच्यते।।

निरुक्ता, न तु सांख्यानामिव जगदुपादानभूता
यच्च पाणिनिप्रभृतिभिः प्रातिपदिकं प्रकृतिरित्युक्तं, तदपि नाम्नो नातिरिच्यते, नातो विभाग (स्य) न्यूनता

तत्र प्रथमोद्दिष्टं नाम लक्षयति–

येन स्वीययदर्थस्य मुख्यतःप्रतिपादने।
स्वोत्तरप्रथमापेक्ष्या तन्नाम स्यात्तदर्थकम्।

यादृशशब्देन स्वोपस्थाप्ययदर्थमुख्यविशेष्यकान्वयबोधार्थं स्वोत्तरप्रथमा विभक्तिरपेक्ष्यते तादृशः शब्द एव तदर्थकं नाम। घटादयो हि शब्दाः प्रथमान्तत्वेन निश्चिता एव स्वार्थमुख्यविशेष्यकं बोधमुत्पादयन्ति नत्वन्यथा, न हि घट इत्यादित इव घटादित्यादितोऽपि घटादिमुख्यकः सुवर्थसंख्यादेरवगमः। पीतपट इत्यादिकः समासोऽपि स्वोत्तरं प्रथमाप्रतिसन्धानदशायामेव पटादिमुख्यधर्मिकस्य पीताद्यन्वयवोधस्य जनकः, अत एव " नापदं प्रयुञ्जीतेति" वृद्धाः। पदं विभक्त्यन्तम्। पीतपटरूपमित्यादावपि नावान्तरवाक्यार्थवोधपूर्वको महावाक्यार्थवोधः, पटादिनाम्नः प्रथमान्तत्वाभावेन तदर्थे पीताद्यन्वयधियः पूर्वमसम्भवात्,विशिष्टवैशिष्ट्यावगाहितायाः प्रमाणान्तरेण पटधर्मिकपीत निश्चयादपि तत्र सम्भवात्, ।न चान्योन्यपरस्परादिनामस्वव्याप्तिः, "एभ्यः प्रथमा नाभिधीयत "इति वृद्धस्मरणादिति वाच्यम्। मल्लाभ्यामन्योन्यं हन्यत इत्यादौ मल्लयोः परस्परकर्तृक- हननकर्मत्वान्वयानुरोधेन तेषामपि प्रथमान्तत्वात्, अन्यथा सकर्मकस्य हन्तेर्भावाख्यातसाकाङ्.-क्षत्वानुपपत्तेरकर्मकस्यैव धातोस्तथात्वात्।

---

* शब्दशक्तिप्रकाशिकात उद्धतमिदं अकरणम् (कारिका - १३)

NOMINALS

The defined stem is of two kinds : *nāman* 'nominal and *dhātu* 'root'. The nominal is not different from what is called a *prātipadika*.

The stem is to be understood as the radical used for etymological interpretation of a word but not as a proximate cause as treated in the *Sāṁkhya* system. What is termed as *prakṛti* by Pāṇini and others is not different from *nāman.*

The first mentioned nomnal is difned as :

"That which expects mainly the first case-termination to present its meaning is the nominal"

The stem which stands in expectancy of the first case-termination which follows it immediately for causing the verbal cognition in which its meaning is projected as the main qualificand is a *nāman* 'nominal'. The words like 'pot' etc. produce the verbal cognition in which its meaning stands prominent as being the nominative determined in the same way and not otherwise. The number i.e. oneness etc. the meaning of the first case suffix in relation to pot which is the prominent, is understood from the expression *ghaṭaḥ* 'pot' itself, but not from the expression *ghaṭāt* which points to fifth case-ending. The compound *pīta-paṭa* 'yellow-cloth' causes the verbal cognition of pot which stands prominent being construed which 'yellow' the qualifier in the state of making out the meaning of the first case-termination immediately following the compound expression. That is why the elders insist upon the fact that "the word without case termination should not be used in the language".

*** This chapter has been taken from śabdaśaktiprakāśikā. V. 13.**

The *pada* is to be understood as a word ending with a case or a verbal termination.

(The opponent argues that) the compound *pīta-paṭa-rūpa* 'yellow-cloth-colour' does not point to the fact that the knowledge of whole compound expression depends on its clause component here. It stands as a qualifier. Since *paṭa* does not contain a first case termination it cannot generate the qualificative verbal cognition taking recourse to the knowledge of *pīta* 'yellow' before hand in the said compound.

The answer is that since this knowledge is a Qualifier qualificand type the constituent *pīta-paṭa* is to be projected as the qualifier-whole which is cognised by another valid means of knowledge (i.e. extra-ordinary perception) prior to the verbal cognition of the whole compound.

(It is also not a point of argument) that the words like *anyonyam* and *parasparam* 'mutual' to which the first case termination is not suffixed cannot be taken on the above line. The elders have attested such case that "from such words the first inflection is not expected". The instance *mallābhyām anyonyaṁ hanyate* 'fighters are beaten by each other' shows that the first inflection has been suffixed to the word *anyonyam*. Since the pattern is a passive kind the word refers to nominative form in relation to the verb 'to kill/beat'. Moreover, the root $\sqrt{han}$ is a transitive verb. Such being the case it relates to objective form which stands as nominative in passive pattern and therefore, the word *anyonyam* stands parallel to nominative form. Further, since the verb $\sqrt{han}$ is transitive there is no possibility of (*bhāva-vācya*) impersonal pattern. Because the impersonal pattern relates to intransitive verb.

-----*-----

नामविभागः

निरूक्तं नाम विभजते-

रूढंच लक्षकञ्चैव योगरूढंच यौगिकम् ।

तच्चतुर्द्धा, परै रूढयौगिकं मन्यतेऽधिकम् ।।

किंचिन्नाम क्वचिदर्थे रूढमेव, यथा गोप्रभृतौ गवादि "गमेर्डोर्दाभाभ्यां नुर्धेट इच्चे" त्यादेरौणादिक(स्य)प्रत्यस्य शक्तिविरहेण यौगिकत्वायोगात् । किंचिल्लक्षकमेव,यथा तीरादौ गङ्गादि, आद्यव्युत्पत्त्या पूरवत्तीरे शक्तिगह-स्याविशिष्टत्वेऽप्याजानिकप्रयोगापत्त्या तदप्रामाण्यकल्पनात्, शक्तेः पूर्वपूर्वप्रयोगनियतत्वात् । किंचिद्योगरूढमेव, यथा पङ्कजाते पदभ्रादौ पंकजादि । किंचिद्यौगिकमेव, यथा पाककर्त्रादौ पाचकादि ।

रूढयौगिकमप्यधिकं नाम, यथा मण्डप महारजतादि, तद्धि कदाचिदवयववृत्त्या योगार्थमेव, कदाच्चिच समुदायशक्त्या रूढ्यर्थमेवाभिधत्त इत्यपरेषाम् मतम् । गौर्वाहिक इत्यादौ तु शक्यार्थसदृशत्वावाच्छिन्नवोधकतया गौणं गवादिपदं गोसदृशादौ लक्षकमेव न तु ततो भिद्यते ।

रूढं नाम लक्षयति च–

रूढं सङ्केतवन्नाम, सैव संज्ञेति कीर्त्यते ।

नैमित्तिकी पारिभाषिक्यौपाधिक्यपि तद्भिदा ।।

यन्नाम यादृशार्थे सङ्केतितमेव न तु यौगिकमपि तद्रढम्, योगरूढन्तु पङ्कजादिकं न तथा । रूढनामैव च संज्ञापदेनाभिलप्यते, न तु रूढादिवत् संज्ञापि नाम्नोऽवान्तरभेदः, येन विभागव्याघातः स्यात् ।

ये तु रूढस्य नाम्नश्चतुर्विधत्वामाहुस्तन्मतमुपन्यस्यति--

जातिद्रव्यगुणस्पन्दैर्धर्मैःसङ्केतवत्तया ।

जातिशब्दादिभेदेन चातुर्विध्यं परे जगुः ।।

---

* शब्दशक्तिप्रकाशिकात उद्धतमिदं प्रकरणम् (कारिका - १६)

गोगवयादीनां गोत्वादिजात्या पश्वाढ्यादीनां लाङ्गूलर्वनादिद्रव्येण धन्यपिशुनादीनां पुण्यद्वेषादिगुणेन चलचपलादीनांच शब्दानां कर्मणावच्छिन्नशक्तिमत्त्वाच्चातुर्विध्यमेव रूढानामिति। यदुक्तं दण्ड्याचार्य्यैः

शब्दैरेव प्रतीयन्ते जातिद्रव्यगुणक्रियाः।

चातुर्विध्यादमीषान्तु शब्द उक्तश्चतुर्विधः।।इति,

तदेतज्जडमूकमूर्खादीनामन्यशून्यादीनाञ्चं शब्दानामपरिग्रहापत्त्या परित्यक्तमस्माभिः।।

CLASSIFICATION OF NOMINALS

A nominal is classified into four categories, viz, *rūḍha* 'conventional' *lakṣaka* 'secondary' *yogarūḍha* 'derivatively conventional' and *yaugika* 'derivative'. Some others also add another kind of nominal termed as *rūḍha-yaugika*' 'conventionally derivative'.

It shows that as far as the meaning is concerned a nominal is somehow conventional. For instance, the nominal *go* 'cow' is a conventional type. This is derived from $\sqrt{gam}$ with the suffix *ḍo*. But it does not point to any import and therefore, cannot be a derivative type. Some are of secondary kind. For instance, *gaṅgā* which is used in the sense of the bank of the river named so. The nominal *gaṅgā* is expressive of river current. When it is used to refer to the bank it does not do the expressive function in the same way as the word *tīra* does to mean the bank. Had it been so it would have been conceived since time immemorial. Moreover, the expressive function of a word is determined by the use being occasioned previously. The nominal *gaṅgā* cannot be used validly in the sense of bank. When it refers to bank it is to be treated as secondary type. Some are derivatively conventional type. For instance, paṅkaja 'lotus' dervatively means 'whatever grows in the mud.' But this word has been conventionally restricted to mean 'lotus' only. Some are derivative. For instance, *pācaka* 'cook'. This has been presented by two constituents namely root and the suffix to mean *pāka-kartā* 'cooking agent'

The conventionally derivative nominals are *maṇḍapa* 'altar' and *mahārajata* 'gold'. These words sometimes present their derivative meaning and sometimes the conventional meaning as a whole, (For instance, *maṇḍapa* derivatively means the man who drinks *scum* (*maṇḍam pibati)* and conventionally it means the altar.)

In case of the instance *gaurvāhika* the nominal *gau* is used to refer to 'the state of similar' to what is expressed as primary meaning with reference to the quality, i.e. idiocy and the like. But it actually refers to what is similar to cow and thus is not different form the secondary type.

The conventional type of nominal is indicated and classifed as follows :-

> "The conventional type of nominal is that which possesses the conventional import. That is termed as *saṁjñā*' 'name'. This is of three types viz. *naimittikī*, 'occasional', *pāribhāṣikī* 'current' and *aupādhikī* 'conditional'."

The name which possess the conventional import and not derivative is a conventional type. The derivatively conventional types such as *paṅkaja* 'lotus' etc. are not like this. The conventional type of nominal is said to be the *saṁjñā*. But it is not a subsidiary kind of name like conventional type by which the classification would be hampered.

A mention is made here as to the views of those who hold that the name is of four kinds :-

> "Others (the Garmmarians) hold that the conventional type of nominal is of four kinds in view of the facts that they are conven-

> tionally expressive of *jāti* 'universal' *dravya* 'substance' *guṇa* 'quality' and *spanda* 'motion' i.e. *kriyā* 'action'".

The words like *go* and *gavaya* are indicated by the universal like *gotva* etc., the words like *paśu* 'beast' and *āḍhya* 'wealthy' by the substance like tail and wealth etc., the words like *dhanya* 'virtuous' and *piśuna* 'malignant' by the quality, merit and malice respectively and the words like *cala* 'moving' and *capala* 'wavering' by the action movement and similar activities.

Thus is said by Acārya Daṇḍin :-

> " Universal, substance, quality and action are expressed by words. On account of such fourfold division the word is said to be of four kind."

This has been discarded by us because the words like *jaḍa* 'idiot' *mūka* 'dumb' *mūrkha* 'uninitiated' *anya* 'different' *śūnya* 'void' and similar other words are not included under such four fold division.

नैमित्तिकी संज्ञा

नैमित्तिकसंज्ञां लक्षयति-

जात्यवच्छिन्नसङ्केतवती नैमित्तिकी मता।
जातिमात्रे हि सङ्केताद्रव्यक्तेर्भानं सुदुष्करम्।।

यन्नाम जात्यवच्छिन्नसङ्केतवत् सा नैमित्तिकी संज्ञा, यथा गोचैत्रादिः। सा हि गौत्वचैत्रत्वादिजात्यवच्छिन्नमेव गवादिकमभिधत्ते न तु गोत्वादिजातिमात्रम्।

पारिभाषिकी रांज्ञा औपाधिकी संज्ञा च

पारिभाषिकीमौपाधिकींच संज्ञां क्रमेण लक्षयति--

उभयावृत्तिधर्मेण संज्ञा स्यात् पारिभाषिकी।
औपाधिकी त्वनुगतोपाधिना या प्रवर्त्तते ।।

उभयावृत्तिधर्मावच्छिन्नसङ्केतवती संज्ञा पारिभाषिकी, यथाऽऽकाश डित्थादि, सा हि द्वितयावृत्तिधर्मेणैव शब्दादिना रूपेण तदाश्रयमभिधत्ते।न चाकाशादिपदस्य गगनादौ निरवच्छिन्नैव शक्तिः, पटादिपदस्यापि पटादौ तादृशशक्त्यापत्तेः।

या चानुगतोपाध्यवच्छिन्ने सङ्केतवती संज्ञा सा त्वौपाधिकी, यथा भूतदूतादिः, सा हि सचेतनावृत्तिविशेषगुणवत्त्ववार्त्ता-हारकत्वाद्यनुगतोपाधिपुरस्कारेणैव प्रवर्त्तते, शब्दादिकन्तु सखण्डत्वेनोपाधिरपि नानुगतो द्वितयावृत्तित्वादतः पारिभाषिके गगनादिपदे नातिप्रसङ्गः। घटत्वादिजातेः संस्थानवृत्तित्वमते यदि घटादिपदं विलक्षणसंस्थानवत्त्वेन शक्तं तदौपाधिकमेव, यदि च परम्परया वैलक्षण्यवत्त्वेन तदा नैमित्तकमेव। वस्तुतः सामान्यस्य शाब्दबुद्धौ स्वरूपतः प्रकारत्वं समवायेनैवेत्यभिहितं प्रागिति।।

' शब्दशक्तिप्रकाशिकात उद्धृ तमिदं प्रकरणम् (कारिका - १६,२२, २३

ये तु जात्यवच्छिन्नसङ्केतवतामपि चैत्रादिपदानां पारिभाषिकत्वमाहुस्तेषां भते त्रैविध्यमन्यथा निर्वक्ति--

यद्वाधुनिकसङ्केत शालि स्यात्पारिभाषिकम् ।
जात्या नैमित्तिकं शक्तमौपाधिकमुपाधिना ।।

यत्रार्थे यन्नामाऽऽधुनिकसङ्केतवत्तदेव तत्र पारिभाषिकम्, यथा पित्रादिभिः पुत्रादौ सङ्केतितं चैत्रादि,यथा वा शास्त्रकृभिः सिद्रध्यभावादौ पक्षतादि। जात्यवच्छिन्नशक्तिमन्नाम नैमित्तिकम्,यथा गो- गवयादि। यदुपाध्यवच्छिन्न शक्तिमन्नाम तदौपाधिकम्, यथाऽऽकाश-पश्वादि।

OCCASIONAL TYPE OF NOMINAL

The occasional type of nominal is mentioned :

> "That which possesses the import delimited by the universal is termed as occassional kind of nominal. Because it is difficult on the part of the word to be expressive of individual if the universal only is admitted to be the import."

That which possesses the import delimited by the universal is termed *naimittikī* 'occasional'. For instance, *go* 'cow' and *caitra* etc. Such a type expresses cow and Caitra etc. being delimited by the universal 'cowness' and the state of being Caitra, i.e. *caitratva* but not only the universal 'cowness' etc.

CURRENT AND CONDITIONAL TYPE OF NOMINAL

The current and conditional type of nominals are indicated respectively

> "That which possesses the property not belonging to both the categories is termed current kind of nominal. That which is characterised by the property belonging to many is called conditional kind of nominal."

That which possesses what is delimited by the property not belonging to both the categories is current kind of nominal. For instance, *ākāśa* 'sky' and *Dittha* the proper name. The nominal *ākāśa* 'sky' is qualified by the property *śabda* 'sound' which belongs to only one category namely *ākāśa* 'sky'. It cannot be said that *ākāśa* is expressive of sky *par se* (without any delimiting determinant). Because the same principle would be applicable to *paṭa* 'cloth' (i.e. *paṭa* 'cloth' would be expressive of cloth *par se* without being qualified by delimiting determinant like 'potness')

That which possesses the import of what is delimited by the general property is *aupādhikī* 'conditional type of nominal'. For instance, *bhūta* 'element' and *dūta* 'messenger'. These are used as being possessed of the characteristics which are conditions like special qualities not belonging to the soul and the state of one who carries a message respectively. Although *śabda* 'sound' likewise a condition being a *sakhanda* property cannot be taken as a property belonging to many. It is, therefore, not over-extended to current kind of nominal.

Some hold that the universal potness means the state of being a peculiar combination of primary parts. To those, if *ghaṭa* 'pot' is regarded to be expressive of what is characterised by the state of being the peculiar combination of primary parts then it is current kind of nominal. If it is regarded separate as delimited by the universal indirectly then it would be occasional kind. As a matter of fact, in knowledge representation the universal is comprehended as a qualifier *par se* through the relation 'inherence'. This has been stated earlier.

Those who hold that the nomial like *caitra* is expressive of what is characterised by *caitratva* 'caitra-ness' to them, it is a current kind of nominal and the three fold division is stated in other way.

> "That which possesses the non-conventional import is current kind. That which are indicated by the universal and the condition, are occasional and conditional repsectively".

The nominal which possesses non-conventional import in point of its meaning is called current kind. For instance, *caitra* stands for a name given to a son by his father. Similarly, the terms *siddhi, abhāva* and *pakṣatā* coined by the authors of *śāstras* are to be taken as non-conventional current kind. That which possesses the conventional import as being delimited by the universal is occasional kind. For instance, *go* and *gavaya*. That which possesses the conventional import as being characterised by the condition is called conditional kind. For instance, *ākāśa* and *paśu* (Which are characterised by sound and tail etc. respectively).

*।। द्रव्यम् ।।

तत्र समवायिकारणं द्रव्यम्। गुणाश्रयो वा। तानि च द्रव्याणि पृथिव्यप्तेजोवाखाकाशकालदिगात्ममनांसि नवैव।

।। गुणाः ।।

अथ गुणा उच्यन्ते। सामान्यवानसमवायिकारणमस्पन्दात्मा गुणः। स च द्रव्याश्रितः। रूप - रस - गन्ध - स्पर्श - क्त्वः –संख्या - परिमाण - पृथक्त्व - संयोग - विभाग - परत्व - अपरत्व - गुरुत्व - द्रवत्व - स्नेह - शब्द - बुद्धि - सुख - दुःख - इच्छा - द्वेष - प्रयत्न - धर्म - अधर्म - संस्कार - भेदाच्चतुर्विंशतिरेव।

।। कर्माणि ।।

इदानीं कर्मोच्यते। चलनात्मकं कर्म। गुण इव द्रव्यमात्रवृत्ति। अविभुपरिमाणेन मूर्त्तत्वापरनाम्ना सहैकार्थसमवेतम्। पूर्वसंयोगनाशे सत्युत्तरसंयोगहेतुश्च। तच्च उत्क्षेपण - अपक्षेपण - आकुंचन - प्रसारणभेदात् पंचधा। भ्रमणादयस्तु गमनेनैव गृह्यन्ते।

---

* तर्क भाषात उद्धृतमिदं प्रकरणम्

* SUBSTANCE

That which is an inherent cause is called substance. It is (also defined as) the substratum of qualities. It is of nine kinds such as : (1) *pṛthivī* 'earth', (2) *vāyu* 'air'' (5) *ākāśa* 'ether' (6) *kāla* 'time' (7) *dik* 'space', (8) *ātman* 'soul' and (9) *manas* 'mind'.

(The inherent cause is difined as that in which the effect inheres when produced. For instance, cloth inheres in the threads).

QUALITIES

The qualities are described as follows :

That which has universal but no motion and serves as a non-inherent cause is called quality. It subsists in a substance. It is of twenty four kinds such as colour, taste, smell, touch, number, dimension, separateness, conjunction, disjunction, remoteness, proximity, heaviness, fluidity, viscidity, sound, intellect, pleasure, pain, desire, aversion, effort, merit, demerit and impression.

(The non-inherent cause is defined as that which adheres to the inherent cause and at the same time, retains its causal capability. For instance, the conjunction of threads is the non-inherent cause of the cloth.

This conjunction being the quality inheres in the treads which are the inherent cause of the cloth. Since it serves as an indispensable antecedent it also fulfils the causal condition. Similarly, the colour of the threads is the non-inherent cause for the colour of the colth.)

*** This chapter has been taken from Tarkabhāṣā**

ACTIONS

Now action is explained as follows :

Action is of the forms of motion and it subsists in substances only like qualities. It co-inheres along with dimensions that donot abide in all pervaling substances. The fact of having such limited dimensions is called *mūrtatva* 'corporeality.' Action is the cause of conjuction at a second place after destruction of its previous conjunction. It is of five kinds :

(i) moving upwards,
(2) moving downwards
(3) contracting,
(4) expanding and
(5) going. Whirling or travelling and such other kinds of action come under 'going'.

सामान्यम्

अनुवृत्तिप्रत्ययहेतुः सामान्यम्। द्रव्यादित्रयवृत्ति, नित्यमेकमनेकानुगतंच। तद् द्विविधं परमपरंच। परं सत्ता बहुविषयत्वात्। सा चानुवृत्तिमात्रहेतुत्वात् सामान्यमात्रम्। अपरं द्रव्यत्वादि अल्पविषयत्वात्। तच्च व्यावृत्तेरपि हेतुत्वात् सामान्यविशेषः।

अत्र कश्चिदाह। व्यक्तिव्यतिरिक्तं सामान्यं नास्ति। अत्र ब्रूमः। किमालम्बना तर्हि भिन्नेषु विलक्षणेषु गोपिण्डेष्वेकाकारा बुद्धिः विना सर्वानुगतमेकं किंचित्। तदेव सामान्यमिति। न नु तस्य व्यावृत्तिकृतैकाकारा बुद्धिरस्ति। तथाहि सर्वेष्वपि गोपिण्डेष्वगोपिण्डेभ्यो व्यावृत्तिरस्ति। तेनाऽगोव्यावृत्तिविषय एवायमेकाकारप्रत्ययोऽनेकेषु गोपिण्डेषु। न तु विधिरूपगोत्वसामान्यविषयः। नैवम्। विधिमुखेनैव एकाकारप्रत्ययस्फुरणात्।

* तर्कभाषात उद्धृतमिदं प्रकरणम्

UNIVERSAL

The universal is that which causes a comprehensive class cognition. It subsists in substances, qualities and actions. It is eternal, one and inheres in many. It is of two types the 'highest' and 'lower'. The highest form is 'existence' as it inheres in many things. This is of the nature of pure universal as it causes a comprehensive cognition. The 'lower' inheres in less number of things such as substance and quality etc. Since it causes both comprehensive and exclusive cognitions it is treated as exclusive universal.

Somebody says here that there is nothing like universal except the individual. Here we say — what is the basis for such comprehensive cognition which arises from the knowledge of many individuals of the particular class such as cow, each having distinct features, if there is no unique unifying element belonging to all individuals of the same class ? This element that unifies is the universal.

(The opponent argues) : Is it not that the comrehensive cognition is due to distinct features of each individual which excludes it from the other? Thus each individual cow is different from all non-cows (like horse and buffalo etc). It is this difference that serves as the comprehensive cognition but not the universal in its positive form.

The answer is as follows : It is not so, because the comprehensive class cognition has always a positive form.

*** This chapter has been taken from Tarkabhāṣā**

विशेषः

विशेषो नित्यद्रव्यवृत्तिः। व्यावृत्तिबुद्धिमात्रहेतुः।

नित्यद्रव्याण्याकाशादीनि पंञ्च, पृथिव्यादयश्चत्वारः परमाणुरूपाश्च।

समवायः

अयुतसिद्धयोः सम्बन्धः समवायः।स चोकतः (कौ पुनरयुतसिद्धौ? ययोर्मध्ये एकमविनश्यदपराश्रितमेवावतिष्ठते तावयुतसिद्धौ। तदुक्तम्-

तावेवायुतसिद्धौ द्वौ विज्ञातव्यौ ययोर्द्वयोः।

अनश्यदेकमपराश्रितमेवावतिष्ठते।।

यथा अवयवावयविनौ, गुणगुणिनौ, क्रियाक्रियावन्तौ, जातिव्यक्ती, विशेषनित्यद्रव्ये चेति। अवयव्यादयो हि यथाक्रममवयवाद्याश्रिता एवावतिष्ठन्तेऽविनश्यन्तः। विनश्यदवस्थास्त्वनाश्रिता एवावतिष्ठन्तेऽवयव्यादयः। यथा तन्तुनाशे सति पटः। यथा वाऽश्रयनाशे सति गुणः। विनश्यत्ता तु नाशकारणसामग्रीसान्निध्यम्।)

नन्ववयवावयविनौ अयुतसिद्धो। तयोःसम्बन्धः समवायः। न चैतद्युक्तम्। अवयवातिरिक्तस्याऽवयविनोऽभावात्। परमाणवो हि यथा तथा सन्निकृष्टा बहवो घटोऽयं घटोऽयमिति गृह्यन्ते। अत्रोच्यते। अस्त्येकः स्थूलो घट इति प्रत्यक्षा बुद्धिः। न च सा परमाणुष्वनेकेष्वस्थूलेष्वतीतिन्द्रयेषु भवितुमर्हति। भ्रान्तेयं बुद्धिरिति चेन्न। बाधकाभावात्।

* तर्कभाषात उद्धृतमिदं प्रकरणम्

PARTICULAR

Particular subsists in eternal substances. It is the cause of exclusive notion of these substances. The eternal substances are five beginning with earth together with the atoms of the four (such as earth, water, light and air.).

INHERENCE

Inherence is a relation between two inseparable things. This has been explained. (What are inseparables ? The things are called inseparable when two things are connected with each other in such a way that one continues to exist so long as the other is not destroyed. This is said , "Those two things are to be known as inseparables if one of them subsists on the other so long as the latter is not destroyed. These are : (1) the parts and the whole, (2) quality and substance, (3) motion and that which moves. (4) the universal and the individual, (5) particular and the eternal substances. The whole etc. clings to the parts so far as they are not destroyed. When the parts begin to perish the whole etc. does not subsist on the other. For instance, when the threads perish the cloth does not exist. When the substratum (i.e. substance) is destroyed its quality is destroyed. The state of being destroyed means presence of all causes for destruction)*

(The opponent, the Buddhist, argues that) you say that the whole and its parts are inseparable and the relation

*** (This passage occurs in the context of explanation of *Karaṇa* in the text, This has been brought to this place for reference as indicated by the author.**

between them is inherence. But this is not correct because there is nothing like whole except its parts. For example, an object is called 'pot' is nothing but an aggregation of many atoms arranged in a particular form. (The answer is as follows) : There is a perception that the pot is a single gross entity.

This cognition does not objectify the atoms which are many, not gross and not perceptible. This cognition cannot be said to be false as there is not sublating judgement which can contradict it.

* अभावः

इदानीं निषेधमुखप्रमाणगम्योऽभावरूपः सप्तमः पदार्थः प्रतिपाद्यते। स चाऽभावः संक्षेपतो द्विविधः, संसर्गाभावोऽन्योन्याभावश्च। संसर्गाभावस्त्रिविधः- प्रागभावः प्रध्वंसाभावोऽत्यन्ताभावश्च।

उत्पत्तेः प्राक् कारणे कार्यस्य अभावः प्रागभावः। यथा तन्तुषु पटाभावः। स चाऽनादिः। उत्पत्तेरभावात्। विनाशी च। कार्यस्य तद्विनाशरूपत्वात्।

उत्पन्नस्य कारणेऽभावः प्रध्वंसाभावः। विनाश इति यावत्। यथा भग्ने घटे कपालयोर्घटाभावः। स च मुद्गरप्रहारादिजन्यः। उत्पत्तिमानप्यविनाशी, नष्टस्य पुनरनुत्पत्तेः।

त्रैकालिकोऽभावोऽत्यन्ताभावः। यथा वायौ रूपाभावः।

अन्योन्याभावस्तु तादात्म्यप्रतियोगिकः। घटः पटो न भवति।

* तर्कभाषात उद्धृतमिदं प्रकरणम्

* ABSENÇE

Now, absence which is cognised as negation is described as the seventh category. This is, in brief, is of two kinds — (1) relational absence and (2) mutual absence. The relational absence is of three kinds (1) antecedent absence (2) subsequent absence or destruction and (3) absolute absence.

Antecedent absence is that which exists in the cause before the effect comes into existence. For instance, absence of cloth in the threads. It has no beginning as it has no origin. But it has an end because the production of effect itself consitutes its end.

Sebsequent absence means destruction which is an absence of an effect in its cause. For instance, absence of a pot in its two halves after it is broken which is brought about by striking it with a hammer. It has an origin but no end as a thing destroyed has no eventuation again.

Absolute absence relates to the absence of all the three times. In other words, an absolute absence is predicated in terms of all times namely past, present and future. For instance, absence of colour in the air.

Mutual absence is that which denies the identity of two things. In other words, it is called difference. For instance, the pot is not cloth.

* This chapter has been taken from Tarkabhāṣā

BIBLIOGRAPHY

1. Primary Sources

1. *Nyāya - darśanam (vātsyāyana-bhāṣya-sahitam)* of Gotam and Vātsyāyana, Edt. Narayaṇ Mishra, Chowkhamba Sanskrit Sansthan, Varanasi, 1983.
2. *Nyāya-mañjarī* of Jayanta Bhaṭṭa, Edt. S.N. Sukla, Chowkhamba Sanskrit Series, Varanasi, 1971.
3. *Nyāya-siddhānt-muktāvalī* of Viśvanātha, Chowkhamba Sanskrit Series, Varanasi, 1972.
4. *śabda-śakti-prakāśikā* of Jagadīśa, Edt. Dhundiraj Shastri, Chowkhamba Sanskrit Sansthan, Varanasi, 1973.
5. *śaktivāda* of Gadādhara Edt. Damodara Shastri, Chowkhamba Sanskrit Series, Varanasi, 1976.
6. *Tarka-bhāṣā* of Keśava Miśra, Chowkhamba Orientalia, Varanasi, 1979.
7. *Tattva-prabhāvalī* of Sri Krishan Vallabhacharya, Chowkhamba Sanskrit Series, Varanasi, 1976.

II. Translations

1. Athalye Y.V. — *Tarka-saṁgraha* of Annaṁ Bhaṭṭa, Bhandarkar Oriental Research Institute, Poona, 1974.
2. Bhattacharya J.V. — *Nyāya-mañjarī,* Part I, Motilal Banarsidass, Delhi, 1978.
3. Chatterjee K.N. — *Word and Meaning — A New Perspective,* Chowkhamba Orientalia, Varanasi, 1980.
4. Chattopadhyaya D.P. & M.K. Gangopadhyay — *Nyāya Philosophy,* Part I, Indian Studies : Past & Present, Calcutta, 1967.
5. ———— — *Nyāya Philosophy*, Part II, Firma KLM Pvt. Ltd. Calcutta, 1981.
6. Iyer S.R. — *Tarka-bhāṣā* of Keśva Miśra, Chowkhamba Orientalia, Varanasi, 1979.

III. ***Secondary Sources***

1. Bātori I.S., et al (Ed.) : *Computational Linguistics,* walter de Gruyter, Berlin, 1989.
2. Charniak E. & Drew MC Dermott — *Introduction to Artificial Intelligence*, Addison-Wesley Publishing company, Canada, 1985.
3. Dash K.C. — *Reference : A Logico Linguistic Identification,* Abhaya House of Publica-

tion, K. Kendrapara, Cuttack - 754211, 1986.

4. Dash K.C. *Relations in Knowledge Representation*, Sri Satguru Publication, Indian Book Centre, Delhi, 1991.

5. Jha, V.N. *The Philosophy of Relations*, Sri Satguru Publications, Indian Books Centre, 1990.

6. ------------- *Viṣayatāvādaḥ*, Publication of C.A.S.S., Class, C, No. 16, University of Poona, 1987.

'7. Mishkoff H.C. *Understanding Artificial Intelligence,* B.P.B. Publications, Delhi, 1986.

8. Page E.S. & L.B. Wilson *Information Representation and Manipulation in a computer,* Affiliated East-West press Pct. Ltd., New Delhi,1978.

9. Tenskey F.N. *Principles of Text Processing,* Affiliated East-West Press, Pvt. Ltd., New Delhi, 1982.

10. Zorkoczy P *Information Technology,* Affiliated East-West Press Pvt. Ltd., New Delhi, 1985.

INDEX